T0275685

RURAL WATER SYSTEMS FOR MULTIPLE USES AND LIVELIHOOD SECURITY

RURAL WATER SYSTEMS FOR MULTIPLE USES AND LIVELIHOOD SECURITY

Edited by

M. DINESH KUMAR
Institute for Resource Analysis and Policy, Hyderabad, India

A.J. JAMES
Institute of Development Studies, Jaipur, Rajasthan, India

YUSUF KABIR
UNICEF Office for Maharashtra, Mumbai, India

ELSEVIER

Amsterdam • Boston • Heidelberg • London • New York • Oxford
Paris • San Diego • San Francisco • Singapore • Sydney • Tokyo

Elsevier
Radarweg 29, PO Box 211, 1000 AE Amsterdam, Netherlands
The Boulevard, Langford Lane, Kidlington, Oxford OX5 1GB, UK
50 Hampshire Street, 5th Floor, Cambridge, MA 02139, USA

British Library Cataloguing-in-Publication Data
A catalogue record for this book is available from the British Library

Library of Congress Cataloging-in-Publication Data
A catalog record for this book is available from the Library of Congress

ISBN: 978-0-12-804132-1

For information on all Elsevier publications visit our web site at https://www.elsevier.com/

Publisher: Candice Janco
Acquisition Editor: Louisa Hutchins
Editorial Project Manager: Rowena Prasad
Production Project Manager: Mohanapriyan Rajendran
Designer: Mark Rogers

Typeset by TNQ Books and Journals

CONTENTS

LIST OF CONTRIBUTORS

S. Bandyopadhyay
Associate Professor, School of Ecology and Environment Studies, Nalanda University, Rajgir, Bihar, India

N. Bassi
Institute for Resource Analysis and Policy, Delhi, India

S. Deshpande
Groundwater Surveys and Development Agency, Pune, Maharashtra, India

J.D. Foster
Chestertown Spy, Chestertown, MD, United States

S. Jagadeesan
Former Managing Director, Sardar Sarovar Narmada Nigam Ltd, Gandhinagar, Gujarat, India

A.J. James
Institute of Development Studies, Jaipur, Rajasthan, India

Y. Kabir
UNICEF Office for Maharashtra, Mumbai, India

D.S. Kumar
Tamil Nadu Agricultural University, Coimbatore, India

M.D. Kumar
Institute for Resource Analysis and Policy, Hyderabad, India

V. Niranjan
Engineering & Research International LLC, Abu Dhabi, United Arab Emirates

R.M. Saleth
Madras School of Economics, Chennai, India

M.V.K. Sivamohan
Institute for Resource Analysis and Policy, Hyderabad, India

N. Vedantam
Engineering & Research International LLC, Abu Dhabi, United Arab Emirates

PREFACE

The idea of a book on rural water systems for multiple uses came towards the end of 2012 when Institute for Resource Analysis and Policy (IRAP)-Hyderabad, UNICEF-Mumbai, and the Groundwater Survey and Development Agency (GSDA)-Pune, Maharashtra, completed a collaborative action research project titled "Multiple Use Water Services to Reduce Poverty and Vulnerability Climate Variability and Change." Several subsequent studies reinforced the finding, which emerged from the Maharashtra study, that many rural water systems—either those meant for irrigation or domestic water supply—serve multiple purposes, depending on the socioeconomic characteristics of the population in the area in which those systems are embedded or its physical environment, defined by hydrology, geo-hydrology, and climate—which is a function of the year and the season. They also showed that the uses of a rural water system can change from year to year, depending on whether it was wet or dry. It can change from the rainy season to summer.

The question was about evolving appropriate technical and socioeconomic considerations for planning and design in order to build resilient multiple-use rural water systems, and crafting institutions and framing policies for their management to meet the domestic and productive water needs, in terms of quantity, quality, and reliability. This required empirical analysis of the performance of different types of multiple-use rural water systems, which are either designed as multiple-use systems or are by default functioning as multiple-use systems, in different physical and socioeconomic settings.

Most of the earlier work on multiple-use water services essentially comprised of theoretical and descriptive work, and lacked empirical analysis of their performance. This book collates experience from different parts of India, representing different agro-ecological and socioeconomic situations, with multiple-use rural water services from different types of socio-technical systems, rural drinking water schemes—single-village and multiple-village schemes, large multipurpose water systems, irrigation tanks—to identify the conditions under which their sustainability is threatened vis-à-vis their ability to meet the domestic and productive water needs of the rural people, which in turn helped us identify the factors that are critical to designing resilient rural water systems. Subsequently, the critical inputs for planning and management of rural water systems for multiple uses, and the institutional and policy regimes that enable enhanced performance are identified.

This volume has compiled, edited and synthesized the works of several scholars from across India, representing several unique physical and socio-economic environments, which are the outcome of several years of thinking or empirical research or practical experience, on improving the performance of rural water systems in order for them to cater to the multiple needs of the people for survival and livelihoods. We are extremely thankful to the contributors for their invaluable works, their timely submissions of manuscripts, and the patience they have shown in responding to the editorial queries.

The chapters (2–11) presented in this volume were outputs of projects undertaken by the respective scholars, either with financial support from external agencies in most cases, or/and the organizations for which they worked, in the rest of the cases. The most prominent among them are the UNICEF-Mumbai office, Sir Ratan Tata Trust-Mumbai, Sardar Sarovar Narmada Nigam Ltd, International Water Management Institute (IWMI), Colombo, GSDA, and IRAP. Chapter "Water, Human Development, Inclusive Growth, and Poverty Alleviation: International Perspectives" is based on empirical research done by its lead author during his association with IWMI during 2007–08, which was extended and built on during his work at IRAP. Four of the subsequent chapters (chapters: Multiple Water Needs of Rural Households: Studies From Three Agro-Ecologies in Maharashtra, Multiple-Use Water Systems for Reducing Household Vulnerability to Water Supply Problems, Sustainability Versus Local Management: Comparative Performance of Rural Water Supply Schemes, and Reducing Vulnerability to Climate Variability: Forecasting Droughts in Vidarbha Region of Maharashtra, Western India) are based on the outcomes of two studies undertaken by IRAP, with financial support from UNICEF, one on multiple-use water services in three regions of Maharashtra (chapters: Multiple Water Needs of Rural Households: Studies From Three Agro-Ecologies in Maharashtra, and Multiple-Use Water Systems for Reducing Household Vulnerability to Water Supply Problems) and the other on development of a decision support tool for drought predictions in Chandrapur district of Maharashtra (chapter: Reducing Vulnerability to Climate Variability: Forecasting Droughts in Vidarbha Region of Maharashtra, Western India), done in collaboration with GSDA, and a third study on designing institutional and policy regimes for sustainable rural water supplies in Maharashtra, undertaken by IRAP independently, with financial support from UNICEF (chapter: Sustainability Versus Local Management: Comparative Performance of Rural Water Supply Schemes).

Chapter "Influence of Climate Variability on Performance of Local Water Bodies: Analysis of Performance of Tanks in Tamil Nadu" on Tamil Nadu tanks was based on a study entitled "Climate Variability and Tank Irrigation Management in Tamil Nadu: An Economic Inquiry into Impacts, Adaptation, Issues and Policies" done under a research grant from the State Planning Commission, Government of Tamil Nadu, Chennai. Chapter "Groundwater Use and Decline in Tank Irrigation? Analysis from Erstwhile Andhra Pradesh" on tanks in Andhra Pradesh was based on a research study undertaken by IRAP, with a small grant from Sir Ratan Tata Trust, Mumbai. The work on safe drinking water was the outcome of a study undertaken by the concerned researcher in Safe Water Network, India. Chapter "Positive Externalities of Surface Irrigation on Farm Wells and Drinking Water Supplies in Large Water Systems: The Case of Sardar Sarovar Project" on positive externalities of canal irrigation on well irrigation and drinking water supplies came out of an SSNNL-sponsored study carried out by IRAP to analyze the social and economic benefits of Sardar Sarovar Project. We are extremely grateful to these organizations for their valuable support and for allowing the authors concerned to contribute the outcomes of those projects to this volume for the benefit of the scholars, academics, and policymakers and development practitioners internationally.

M.D. Kumar
Y. Kabir
A.J. James
Editors

ACKNOWLEDGMENTS

The editors are thankful to the five reviewers, who have gone through the book proposal, and the editorial office of Elsevier Science, who have gone through the sample chapters, for offering invaluable comments and suggestions, which have immensely helped in framing the questions which the volume addresses, and enhancing the value of this volume. We sincerely hope the current volume titled *Rural Water Systems for Multiple Uses and Livelihood Security* will raise the level of international debate on building sustainable rural water systems for domestic and productive needs in developing countries, by enabling them to factor in the key physical and socio-economic, institutional and policy factors that characterize water resource development, use, and management in the rural areas of those countries, in deciding on the quantity and quality of water, operational rules, and technical infrastructure for the options being considered.

CHAPTER 1

Introduction

M.D. Kumar
Institute for Resource Analysis and Policy, Hyderabad, India

Y. Kabir
UNICEF Field Office, Mumbai, Maharashtra, India

A.J. James
Institute of Development Studies, Jaipur, India

1.1 CONTEXT

A large proportion of the world's population are without access to safe water for drinking lives in India. In spite of huge public investments in water supply, India has made limited progress in improving access to safe water in rural areas in terms of physical access, quality, adequacy, reliability, and dependability. While the Economic Survey of 2013–14 found that more than 80% of the rural population has access to safe water for drinking through taps, bore wells, tube wells, or hand pumps (MoHA, 2014), only a little over 30% of households had access to "tap water" in their dwelling premises as per the 2011 Census. In addition, "access" to supply, as it is defined today, requires only physical proximity to water infrastructure—regardless of whether it provides adequate amount of water or not (see, WHO/UNICEF, 2011). Therefore, these figures, while hiding considerable regional and local variations in drinking water availability, endorse the anecdotal evidence of women walking for miles during summer months in semiarid India to fetch water and having to dig river beds to strain pitiful amounts of drinking water.

This situation is in sharp contrast to the growing water use in irrigation, with the country becoming the largest user of irrigation water in the world, with around 97 m ha of crop land being watered (Kumar et al., 2012a). Planning and execution of modern rural water supply systems without due consideration to the physical and environmental sustainability of the resource base, real water needs of rural households, the actual cost of operation and maintenance of the system, and of the institutional capacities required to manage the systems, are major reasons for this dichotomy.

Rural Water Systems for Multiple Uses and Livelihood Security
ISBN 978-0-12-804132-1
http://dx.doi.org/10.1016/B978-0-12-804132-1.00001-9

Rural water supply sources often become defunct, or their supply levels drop or the quality of water from these sources deteriorates, with the result that there is heavy slippage.

With the advent of reforms in the rural water and sanitation sectors, the past one-and-a-half decades have seen governments overemphasizing decentralized schemes. "To enable the community to plan, implement and manage their own water supply systems," as part of the paradigm shift, the government of India wants the states to transfer the program to the Panchayati Raj Institutions (PRIs), particularly to the Gram Panchayats for the management of water supply within the village (MoRD, 2010). Hence, the focus has been on hand pumps and small groundwater-based village water supply schemes that the Panchayats could run. The fact that water scarcity concerns problems with allocation more than overall physical availability of water and that there are competing demands for water in rural areas is hardly taken into consideration while promoting such schemes. Hence, these schemes have poor dependability, and failure of schemes due to drying up of sources is also rampant. There is little recognition of the fact that rural communities, particularly poor rural households, have multiple water needs—domestic and productive.

Traditional water supply systems such as tanks, ponds, open wells, and lakes from a variety of physical and environmental settings—from humid and subhumid subtropical regions to arid tropics—are used to meet the multiple water needs of rural India, including drinking and cooking, water for personal hygiene (washing, bathing, sanitation), livestock drinking, and recreation (Agarwal and Narain, 1997). Amongst these are tanks, built centuries ago by *zamindars* (feudal landlords) and kings, and by the British Government in India during the 18th and 19th centuries. They are concentrated in the southern states of Andhra Pradesh, Karnataka, and Tamil Nadu, where they provided a major source of irrigation until the mid-1960s. But these water bodies are fast degrading as a result of a variety of physical, socioeconomic, and institutional changes and the vast majority of them have gone into disuse. The declining hydrological performance of the tanks, and the implications for irrigation and associated consequences for their multiple functions from a hydrological perspective have not been thoroughly examined. Instead, an attempt has been made to attribute the declining performance of tanks to the erosion of community management structures, which were responsible for their upkeep, and subsequent management takeover by the government (eg, Pradan, 1996; Rao, 1998); the lack of interest of command area farmers in managing these common pool resources (von Oppen and Rao, 1987; Shankari, 1991; Sekar and Palanisami,

2000; Balasubramaniyam and Bromley, 2002); and to the development approach followed during the British rule centered on modern large irrigation systems with the decline of tanks (Paranjape et al., 2008).

Mosse (1999), however, challenged the long-held view of scholars working on tanks in South India that the collapse of community institutions was the major cause of decline of tanks and contends that, even in the past, communities made little investment in the upkeep of tanks. According to him, it was the *zamindars* and kings who not only built most of the tanks but also spent money on their upkeep and the fall of the institution of overlords led to the decline of the tanks.

Researchers and civil society organizations, however, continue to stress that traditional tanks are sustainable water supply systems, and the institutions built around them are superior to modern water institutions in terms of their effectiveness in equitable water allocation and sound water management (eg, Sakthivadivel et al., 2004; ADB, 2006), barring the exceptions of Shah (2008)[1]. In parallel, there is increasing emphasis from governments, prompted by World Bank–supported tank rehabilitation projects in Karnataka and Andhra Pradesh, to look at the revival of traditional water systems as a substitute to building modern water systems—which involve construction of large reservoirs and canal systems—and create institutions for their management. This also suited the changing sociopolitical context, wherein the opposition to building of large water systems on social and environmental grounds grew, both nationally and internationally. Particularly, as noted by Jagadeesan and Kumar (2015), large irrigation systems based on reservoirs and canals were considered inefficient and ecologically unsustainable by social and environmental activists.

This bias against large systems and the sensitiveness of the issues associated with building large reservoirs also led to the overdependence on groundwater as a sustainable source of irrigation and drinking water supplies.

Supported by government policies, well irrigation expanded rapidly in many semiarid regions, often at the expense of the environmental and social needs of the local communities. These regions, which experienced intensive groundwater extraction for irrigation (Kumar, 2007; Kumar et al., 2012a),

[1] Shah (2008) argued that tanks symbolized an increasingly extractive statecraft involving coerced labor; the expropriation of surplus by elites; and the spread of technological choices that could be environmentally unsound and that often resulted in forced displacement, uncertainty, technological vulnerability, and social anxiety and violence. Further, tanks as techno-sociological artifacts were socially embedded in societies and economies that were organized for warfare, and that sustained sharp social hierarchies, and were often violent to women and people from lower castes (Shah, 2008, p 673).

often suffered a severe shortage of water for drinking during summer months, as groundwater-based public drinking water sources (such as wells, tube wells, and hand pumps) dried up due to severe competition from irrigators tapping the same aquifers. Water from deep aquifers suffered from poor chemical quality (TDS and fluorides). This forced water supply agencies in some states to look for more sustainable and dependable sources of water to meet basic needs.

Despite the growing challenge of managing sustainable water supplies from groundwater-based sources, many water supply agencies and civil society groups continue to extol groundwater-based drinking water sources for virtues such as decentralized management by local communities, low cost, and equity in access to water, as against surface water-based regional water supply schemes, often criticized for their high capital and maintenance costs and poor inter-village equity in water distribution. However, apart from achieving decentralized management, concerns of intravillage equity are yet not addressed in local groundwater-based drinking water supply schemes, as the poor people in rural areas suffer the most when drinking water sources dry up during summer months, while rich households manage to access water from their own private water sources.

In spite of recognizing the severe competition that the drinking water sector faces from larger water-consuming sectors, public water utilities, which use modern water development and distribution technologies, generally design rural water supply schemes as *single-use systems* to meet only domestic water needs. The design of these schemes does not consider the needs of rural communities for sustainable livelihoods, and hence fails to support these communities in performing economic activities using the water supplied from such systems. As a result, these communities show a low level of willingness to pay for the water supply which, in turn, affects the sustainability of the systems as official agencies are not able to recover the costs of their operation and maintenance. Thus a vicious circle is perpetuated.

1.2 RATIONALE FOR THE BOOK

Most developing countries had their own traditional rural water systems before modern water supply technologies were introduced. Tanks, ponds, and lakes are traditional rural water systems in India. There is a vast literature on the multiple-use benefits of wetlands, particularly tanks and lakes. Arguments are made for renovation of these water bodies for meeting multiple water needs of village communities (Paranjape et al., 2008). The government and the donors are investing billions of dollars every year in India, particularly in the southern states, to rehabilitate irrigation tanks to enhance their performance (ADB, 2006). The cost of rehabilitation was as high as Rs. 56,500 per ha for the World Bank–funded Karnataka JSYS (*Jala Samvardhane Yojana Sangha*) project.

The decision to invest in tank rehabilitation, however, has not yet been backed by any analysis of the hydrological factors which affect their performance. Instead, their poor performance is more often attributed to the collapse of tank management institutions. In fact, various physical and socioeconomic processes alter the hydrology of tanks in rural settings, thereby affecting their performance, and hence there is a need to evolve quantitative criteria to identify tanks that are suitable for rehabilitation so as to optimally utilize scarce financial resources.

There is, unfortunately, a conspicuous paucity of empirical analysis quantifying the various benefits of tank rehabilitation, how these benefits change with climatic variability, and how maximizing one type of benefit can affect others—issues that are critical for rational investment decision-making on renovation. Hence, there is an urgent need to: quantify these benefits from different agro ecological systems; and understand the economic and environmental tradeoffs in optimizing these benefits.

As regards modern rural water systems, there is abundant theoretical discourse on designing village water supply schemes as multiple-use systems (see inter alia, van Koppen et al., 2009). While they concern developing countries, they are quite generic and not country- or region-specific. Also, there is very little understanding of the type of water needs in rural households; how these needs vary across socioeconomic and climatic settings; the extent to which these needs are met by the existing single-use water supply schemes; and what coping mechanisms are used to address unmet demands. More empirical assessments are needed to design multiple-use water systems—including analysis of how the multiple water needs of rural communities change with agro ecology, the type of water needs for which rural communities depend on public water supply schemes; and the alternate sources used to meet various different domestic and productive water needs over different seasons in a single year.

Since there are not many multiple-use water systems (MUWSs) in existence by design, it would help rural water system managers to understand how existing public water supply schemes in rural areas can be retrofitted to augment their supply potential and to improve their dependability, and thus make them sustainable sources of water supply for domestic and productive needs. It is important to document the process of developing technical designs of multiple-use water systems for different agro ecological and socioeconomic settings. The design considerations are extremely important, and, obviously, have to be based on extensive review and synthesis of multiple-use water systems around the world, in the absence of evidence from well-functioning MUWSs in India. The cases of well-functioning MUWSs from South Asia, South-East Asia, Latin America, North and East Africa have been documented by scholars in the recent past and should be critically analyzed for use in India.

Large water systems often produce multiple-use benefits indirectly—such as through recharge to groundwater or replenishment of local wetlands, in the form of improved water supplies from existing wells, dilution of minerals in groundwater, raised water table reducing the cost of energy for pumping groundwater, and improved sustainability of well irrigation in command areas, and therefore serve as multiple-use systems. But, there is paucity of literature providing quantification of such indirect benefits from large water systems (Jagadeesan and Kumar, 2015). Conventional methodologies, which evaluate the economic benefits from large water systems, are inadequate for these and innovative methodologies are required to assess their social and environmental costs and benefits. Such methodologies need to be employed for real-life projects to demonstrate their usefulness and to estimate the relative magnitudes of indirect and direct benefits.

A nuanced and clear understanding of all these issues—the nature and relative size of multiple water needs of rural communities; how these needs change in quantitative terms across agro climatic and socioeconomic settings; what multiple benefits the traditional rural water systems produce; how these benefits vary with climate; under what conditions these systems can be rehabilitated for enhancing their benefits; how modern water systems can be designed and built as multiple-use water systems; what indirect social, economic, and environmental benefits the large modern water systems produce; and how to maximize those benefits for them to become multiple-use water systems, etc.—is necessary to design and build resilient and sustainable rural multiple-use water systems.

1.3 A NORMATIVE FRAMEWORK FOR ANALYZING THE PERFORMANCE OF MULTIPLE-USE WATER SYSTEMS

All water systems can potentially serve multiple purposes, be it in rural areas or urban areas, as water itself has multiple roles as an economic good, a social good, and an environmental good. Unlike in developed countries, water supply services to different sectors such as irrigation, domestic water supply, livestock water supply, fisheries, and recreation, in developing countries are still not fully formalized, though the number of formal water supply systems, especially in irrigation and domestic water supply, is on the rise. There is a general tendency among rural populations in these countries, therefore, to meet water needs of one particular "use" from sources meant for another use. For instance, village women use running canals to wash clothes and clean vessels, and rural people in general use canal water also for bathing. The livestock-rearing community would take their animals to running canals, for feeding and often for washing them.

Rural domestic water supply systems such as hand pumps on bore wells, which are designed to meet basic survival needs (say at the rate of 40 L per capita per day), are also used to clean and water livestock and for a range of productive uses, including tea-making, kitchen gardens, vermicomposting units, and pottery (Lovell, 2000; Moriarty et al., 2004; James, 2004). In certain other situations, if the water from the source is not adequate to meet the productive water needs, which is a priority for survival and livelihoods of poor rural households, the households might spend time fetching water from other sources.

In view of the change in pattern of use from the original design, the performance of these systems needs to be assessed vis-à-vis the different types of benefits *actually* realized, rather than the *intended* benefits.

Actual benefits, however, would depend entirely on the local situation. As van Koppen et al. (2009) points out, it can result in either households not being able to realize the full potential of water as a social good or in the damage of the system. For instance, in the first case, if the water from the drinking water source is of good quality but just adequate for human consumption in the village/hamlet, then diversions for livestock and other productive uses could reduce welfare benefits: The reason is drinking water demand has the highest priority, and if such diversions deprive other households of water needed for basic survival needs, the full benefits of the "social good," ie, water supply, will not be realized. Similarly, in the second case, in the stretches of canal downstream of where clothes are washed, the water will be polluted and unsuitable for livestock drinking because of the presence of detergents. Further, the continued use of canal water for washing, bathing, and animal washing could damage the canal lining, unless steps are provided for people to climb down to the water.

Households with individual water supply connections may also divert water for kitchen gardens. This would require an additional pipeline which takes water to the backyard, over and above the capital investment (for retrofitting) to get extra water from the source or from a new source. However, such diversion would be economically feasible only if the cost of supplying the additional water from the drinking water source for the kitchen garden (including the cost of additional infrastructure) is *less* than the cost of setting up separate infrastructure to get the same amount of water of irrigable quality.

Hence, a rural water system designed as a single-use system and converted into a "multiple-use system" can be said to be performing well only if the total value of the net social, economic, and environmental benefits produced from such uses exceeds the value of the net benefits from its alternative (previous or current) use.[2] Over and above, the process of conversion of the single-use system into a multiple-use one should not result in any compromise on the "existing uses" in terms of quantity and quality of water accessed and the ease with which it is accessed by different stakeholders. Therefore, any rural water system originally designed as a single-use system and now being converted into a multiple-use system should not only ensure (continued) equity of access to all stakeholders and (continued) quality of physical infrastructure but should also enhance the benefits received (including social and environmental benefits).

Performance assessment of systems designed to be "multiple-use water systems" should be based on strong economic considerations that assess the cost of production and supply of water to meet multiple water needs against the (sum of) different benefits generated—provided service levels (in terms of quantity, quality, reliability, and access) are at least as much as in "single-use systems." Additionally, however, the quality, reliability, and access offered by the system should be flexible enough to change according to the type of water needs. Building a multiple-use system to cater to all water needs of a particular area, instead of a (set of) single-use systems that together cater to those needs, would be justified only if the potential net benefits (measured through an environmental and social cost-benefit analysis) of the former outweigh those of the latter.

Such assessments should also be done across different years, as the performance of most traditional multiple-use systems varies across years according to climatic conditions, changes in nature, and the quantum of benefits. A good example is a tank system, which receives its inflows from local catchments and caters to local demands (see Kumar et al., 2012b). Irrigation

[2] In the case described above, the cost of retrofitting should also be deducted from the benefits while evaluating the net benefits, along with the cost of the old infrastructure.

tanks in semiarid and arid regions with low to medium rainfall, experiencing high variability in rainfall and other weather parameters, perform well during high rainfall (wet) years, with large inflows, resulting in larger area under irrigated crop production, larger volume of fish production, and larger population receiving benefits such as water for domestic and livestock uses. In drought years, however, inflows reduce drastically, with a proportional shrinking of the area under irrigated crop production and fish production, and with a smaller proportion of people receiving domestic and livestock water supply benefits. Drought years would nevertheless find some new uses of the tank system, such as crop cultivation in the tank bed or the use of silt from the tank bed as manure for crop land. In high rainfall regions, in contrast, the change from a wet year to a drought year is just the opposite and higher economic values are realized in *drought* years as irrigation becomes critical during such years (Kumar et al., 2013). Hence, the benefits from the tank depend on which year is considered for assessment and under which climatic regime. The performance assessment therefore should be done temporally, covering wet, dry, and normal years.

1.4 SCOPE OF THE BOOK

This book is about technological, institutional, and policy choices for building rural water supply systems in developing countries like India, which are sustainable from physical, economic, and ecological points of view and which address the water security and livelihood concerns of poor rural communities. The analysis presented in chapter "Water, Human Development, Inclusive Growth, and Poverty Alleviation: International Perspectives" uses global data sets on water security (sustainable water use index), progress in human development (human development index), economic conditions (per capita GDP levels), income inequality, and human poverty (human poverty index) of nations to show that improving the security of water for all social, economic, and environmental uses improves economic conditions of nations, through the human development route with a positive impact on poverty reduction and income inequality.

The detailed empirical analysis presented in chapter "Multiple Water Needs of Rural Households: Studies From Three Agro-Ecologies in Maharashtra" tests the hypothesis that rural people have multiple household-level water needs that are both domestic and productive. It also argues that decentralized public water supply schemes that tap local groundwater are inadequate to meet these needs and, as a result, rural communities are forced to depend on multiple sources, spending significant amounts of time and labor in the process. Many groundwater-based rural water supply schemes, especially in hard rock areas, fail due to poor yield resulting from resource depletion and water quality deterioration.

The cost norms for building rural water supply schemes followed by the state governments do not reflect the full cost of these schemes, which includes capital costs, costs of operation and maintenance (O&M), and capital maintenance costs. Investments in rural water supply systems are almost exclusively on capital expenditure for infrastructure, while components such as planning and designing, capital maintenance, and water quality receive little or no allocation, with the result that the actual life of infrastructure is much less than the "normative life" used as the basis for cost estimates (Reddy and Batchelor, 2012). A "life cycle cost" approach for investment decisions in water supply infrastructure could thus favor investments in technologies that are more sustainable, rather than those which appear to be "least cost," purely in terms of capital expenditure.

There is growing appreciation that conversion of "single-use systems" for water into multiple-use systems (without causing much damage to the physical infrastructure) improves all four dimensions of livelihoods related to water, namely, freedom from drudgery; health; food production; and income. In the context of rural water supply schemes, it is less well understood that even a *marginal* improvement in drinking water supply infrastructure through retrofitting and a *marginal* increase in the volume of water supplied could result in a remarkable increase in the social and economic value of the water supplied. But, it is often believed that additional water resources to meet multiple needs would be difficult to find in rural areas and, therefore, rural water supply systems are designed to meet the bare minimum needs of the households.

Chapter "Multiple Use Water Systems for Reducing Household Vulnerability to Water Supply Problems" presents the techno-institutional model for multiple-use water systems (MUWS), based on analysis of different agro ecological and socioeconomic settings (from high-rainfall, hilly areas to low-rainfall, drought-prone areas) in Maharashtra. The analysis shows that it is possible to augment the supplies of public water supply systems, even in the most water-scarce regions, through the right choice of technologies and thus to meet the multiple water needs of rural households even in peak summer months. It also describes the kind of retrofitting required in existing public systems and the institutional model for implementing MUWSs in the village context. It also shows that, if implemented effectively, MUWSs can reduce the vulnerability of rural households to problems associated with inadequate water supplies for domestic and productive needs during summer, which become severe during droughts (Kabir et al., 2015). In the context of irrigation tanks, earlier analysis for western Odisha had shown what technical improvements would be required to convert public water systems into MUWSs that could cater to the domestic as well as the productive water needs of rural communities (Kumar et al., 2013).

Using an empirical study of different types of rural water supply schemes, chapter "Sustainability Versus Local Management: Comparative Performance of Rural Water Supply Schemes" shows that schemes built with the sole objective of promoting decentralized governance and management of water supply favor a certain techno-institutional model of water supply, but fail to meet dependability and sustainability criteria. In contrast, techno-institutional models, which are more dependable, and which meet the criterion of sustainability, are not preferred by local communities, due to lack of techno-managerial capacity to operate and maintain them. The chapter, based on the management performance of different types of water supply schemes—from single village schemes based on groundwater to regional schemes based on surface reservoirs—explodes the myth that regional water supply schemes are technically unviable and unsustainable from an economic perspective.

Case studies from two different agro ecologies from south Indian peninsula, namely, Tamil Nadu and erstwhile Andhra Pradesh, presented in chapters "Influence of Climate Variability on Performance of Local Water Bodies: Analysis of Performance of Tanks in Tamil Nadu" and "Groundwater Use and Decline in Tank Irrigation? Analysis From Erstwhile Andhra Pradesh", respectively, show that well-maintained tanks that are primarily used for irrigation serve multiple water needs of rural households (such as irrigation, domestic water use, livestock water use, and recreation) and also serve ecological purposes such as groundwater recharge. The study from Tamil Nadu also examines how these multiple functions performed by the tanks change with climate variability—essential analysis to create or assess drought preparedness and mitigation strategies. Efforts to *revive* traditional water systems such as tanks have focused solely on engineering measures of capacity enhancement (desilting, embankment stabilization), and creating water users' associations for managing them. The study from Andhra Pradesh presented in chapter "Groundwater Use and Decline in Tank Irrigation? Analysis From Erstwhile Andhra Pradesh" shows that the measures such as desilting, cleaning of supply channels, and clearance of catchment, which are not considered the real factors responsible for the degradation of the original system, may only have limited success and only result in a waste of public money. It argues that hydrological and socioeconomic characteristics of the catchment are critical factors that determine the success of rehabilitation efforts.

While droughts magnify problems of water insecurity (and more so in hard rock areas with poor groundwater endowment), predicting this phenomenon is extremely important for mitigating their societal impacts. Chapter "Reducing Vulnerability to Climate Variability: Forecasting Droughts in Vidarbha Region of Maharashtra, Western India" provides a methodology for predicting meteorological droughts, the likely impacts of rainfall variability on groundwater resource availability (water level fluctuations and recharge) and the drinking water situation, in hard rock regions, which are also prone to climate variability and droughts. It demonstrates the methodology for a drought-prone district in Maharashtra by analyzing the probability and frequency of occurrence of meteorological droughts (and the probable intensity), how these droughts impact water level changes in wells in hard rock regions in different seasons, and the utilizable monsoon recharge of groundwater, the only source of water in hard rock aquifers for such regions. The model also predicts outcomes in terms of changes in cropping intensity and water levels in wells during summer months in such areas.

Good-quality water for drinking is a continuing public health challenge in many semiarid and arid parts of India, but with shifts over time. While pathogenic contamination of surface water and water from shallow aquifers was sought to be addressed through massive tube-well drilling programs in the 1980s and 1990s, chemical contamination of water in aquifers—with fluorides, chlorides, nitrates, bicarbonates, arsenic, and TDS—is now a widespread problem, often causing irreversible public health damage, and is forcing yet another rethink.

Access to safe drinking water is largely seen as a government responsibility. Though government policies are still focused on increasingly scarce "alternative safe sources," there is growing realization of the need to purify locally available poor-quality water for domestic consumption in rural areas. Large-scale rural drinking water schemes with better-quality, dependable and sustainable regional sources have not materialized much in the water-scarce regions of India, barring Gujarat (although an ambitious project to cover the entire rural population through a piped water supply is in the pipeline in the newly formed state of Telangana). Decentralized solutions face challenges of appropriate technology, management capacity, financing options, and environmental impacts and, under these circumstances, models of public–private partnerships, community-managed systems, and social enterprises have emerged. Chapter "Sustainable Access to Treated Drinking Water in Rural India" explores these models with the help of case studies, to understand what needs to be done, and by whom, for a sustainable and scalable solution.

Starting with a review of global experience, chapter "Positive Externalities of Surface Irrigation on Farm Wells and Drinking Water Supplies in Large Water Systems: The Case of Sardar Sarovar Project" uses the analysis of a large, multipurpose water resources project in India—the Sardar Sarovar Project, built to meet the irrigation, domestic, and industrial water needs in the arid and semiarid regions of Gujarat—to illustrate the role of large surface water systems in ensuring a sustainable water supply in regions which experience high climatic variability and severe droughts, and where the irrigation and drinking water sources based on groundwater are highly unsustainable. The analysis shows how gravity irrigation from the project induced impacts that go far beyond the intended benefits of enhanced agricultural production for canal water users. It created benefits of improved sustainability of drinking water sources in rural and urban areas and thereby reduced the cost of water supply, improved well yields and lowered the incidence of well failures and reduced energy use in groundwater pumping, resulting in higher income for well irrigators, and increased wage rates for farm laborers in canal-irrigated areas.

This book illustrates the multidimensional nature of the strategies that are required to build resilient and sustainable rural water supply systems—in contrast to the limited approaches that suggest either partial, economically unviable, institutionally weak and technology-driven approaches or those that seek a return to a "glorious" but unattainable past. Chapter "Re-Imagining the Future: Experiencing Sustained Drinking Water for All" re-imagines what a resilient and sustainable system may look like, and suggests indirectly how it may be achieved.

Much money has already been wasted by governments of developing countries in water supply schemes that become non-functional in a few years of operation, especially when we consider the fact that these countries face a paucity of resources to fund such schemes. The knowledge, experience, and suggestions presented here could be vital for developing economies that are looking for a clear, multidimensional and sustainable vision of the ultimate goal of providing adequate, clean and sustainable rural water supply services.

REFERENCES

Agarwal, A., Narain, S. (Eds.), 1997. Dying Wisdom: Rise, Fall and Potential of Traditional Water Harvesting Systems in India. Centre for Science and Environment, New Delhi. 4th Citizen's Report.

Asian Development Bank, 2006. Rehabilitation and Management of Tanks in India: A Study of Select States. Asian Development Bank, Philippines.

Balasubramanian, R., Bromley, D.W., 2002. Mobilizing indigenous capacity: A portfolio approach to rehabilitating irrigation tanks in South India. University of Wisconsin Madison, Wisconsin. www.aae.wisc.edu/www/events/papers/balasubramanian.pdf.

Jagadeesan, S., Kumar, M.D., 2015. The Sardar Sarovar Project: Assessing Economic and Social Impacts. Sage Publications, New Delhi.

James, A.J., 2004. Linking water supply and rural enterprise: issues and illustrations from India. In: Moriarty, P., Butterworth, J., van Koppen, B. (Eds.), Beyond Domestic: Case Studies on Poverty and Productive Uses of Water at the Household Level. IRC International Water and Sanitation Centre, Delft, The Netherlands. Available at: www.irc.nl/page/6129.

Kabir, Y., Vedantam, N., Kumar, M.D., 2015. Women and water: vulnerability from water shortages. In: Cronin, A., Mehta, P.K., Prakash, A. (Eds.), Gender Issues in Water and Sanitation Programmes: Lessons from India. Sage Publications, New Delhi, p. 45.

van Koppen, B., Smits, S., Moriarty, P., Penning de Vries, F., Mikhail, M., Boelee, E., 2009. Climbing the Water Ladder: Multiple-Use Water Services for Poverty Reduction. TP series no. 52. IRC International Water and Sanitation Centre and International Water Management Institute, Hague, The Netherlands.

Kumar, M.D., Sivamohan, M.V.K., Narayanamoorthy, A., 2012a. The food security challenge of the food-land-water nexus in India. Food Security 4 (4), 539–556.

Kumar, M.D., Vedantam, N., Bassi, N., Puri, S., Sivamohan, M.V.K., 2012b. Making Rehabilitation Work: Protocols for Improving Performance of Irrigation Tanks in Andhra Pradesh. Institute for Resource Analysis and Policy, Hyderabad.

Kumar, M.D., Panda, R., Vedantam, N., Bassi, N., 2013. Technology choices and institutions for improving economic and livelihood benefits from multiple use tanks in western Odisha. In: Kumar, M.D., Sivamohan, M.V.K., Bassi, N. (Eds.), Water Management, Food Security and Sustainable Agriculture in Developing Economies. Routledge, London, UK, pp. 138–163.

Kumar, M.D., 2007. Groundwater Management in India: Physical, Institutional and Policy Alternatives. Sage Publications, New Delhi.

Lovell, C., 2000. Productive Water Points in Dry Land Areas: Guidelines on Integrated Planning for Rural Water Supply. ITDG Publishing, London, UK.

Ministry of Home Affairs, 2014. Economic Survey, 2013-14. Office of the Registrar General, Ministry of Home Affairs. Government of India, New Delhi.

Ministry of Rural Development, 2010. Movement towards Ensuring People's Drinking Water Security in Rural India Framework for Implementation. Rajiv Gandhi National Drinking Water Mission, National Rural Drinking Water Programme, Department of Drinking Water Supply, Ministry of Rural Development, Government of India, New Delhi.

Moriarty, P., Butterworth, J., van Koppen, B. (Eds.), 2004. Beyond Domestic: Case Studies on Poverty and Productive Uses of Water at the Household Level, Technical Paper Series No. 41. IRC International Water and Sanitation Centre, Delft, The Netherlands. Technical Paper no. 41.

Mosse, David, 1999. Colonial and contemporary ideologies of 'community management': the case of tank irrigation development in south India. Modern Asian Studies 33 (2), 303–338.

von Oppen, M., Rao, K.S., 1987. Tank Irrigation in Semi-arid Tropical India: Economic Evaluation and Alternatives for Improvement. Research Bulletin no. 10. International Crops Research Institute for the Semi-Arid Tropics, Patancheru, Andhra Pradesh.

Paranjape, S., Joy, K.J., Manasi, S., Latha, N., 2008. IWRM and traditional systems: Tanks in the Tungabhadra system. STRIVER policy brief no. 4. NIVA/Bioforsk, Norway.

Pradan, 1996. Resource management of minor irrigation tanks and Panchyati Raj. In: Paper Presented at the Seminar on Conservation and Development of Tank Irrigation for Livelihood Promotion, Madurai, Tamil Nadu, India.

Rao, G.B., 1998. Harvesting water: irrigation tanks in Anantapur. Wasteland News 13 (3).

Reddy, V.R., Batchelor, C., 2012. Cost of providing sustainable water, sanitation and hygiene (WASH) services: an initial assessment of a life-cycle cost approach (LCCA) in rural Andhra Pradesh, India. Water Policy 14 (3), 409–429.

Sakthivadivel, R., Gomathinayagam, P., Shah, T., 2004. Rejuvenating irrigation tanks through local institutions. Economic and Political Weekly 39 (31), 3521–3526.

Sekar, I., Palanisami, K., 2000. Modernised rain fed tanks in south India. Productivity 41 (3), 444–448.

Shah, Esha, 2008. Telling otherwise a historical anthropology of tank irrigation technology in South India. Technology and Culture 49 (3), 652–674.

Shankari, U., 1991. Tanks: major problems in minor irrigation. Economic and Political Weekly 26 (39), A115–A124.

Wold Health Organization/UNICEF, 2011. Drinking Water Equity, Safety and Sustainability: JMP Thematic Report on Drinking Water 2011. World Health Organization and UNICEF, Geneva.

CHAPTER 2

Water, Human Development, Inclusive Growth, and Poverty Alleviation: International Perspectives

M.D. Kumar
Institute for Resource Analysis and Policy, Hyderabad, India

R.M. Saleth
Madras School of Economics, Chennai, India

J.D. Foster
Chestertown Spy, Chestertown, MD, United States

V. Niranjan
Engineering & Research International LLC, Abu Dhabi, United Arab Emirates

M.V.K. Sivamohan
Institute for Resource Analysis and Policy, Hyderabad, India

2.1 INTRODUCTION

As water scarcity hits many developing regions of the world, there is now a renewed interest in understanding how growing threats to water security affect future progress in human development and economic growth of nations (Grey and Sadoff, 2005; Grey and Sadoff, 2007). The international development debate is, however, heavily polarized between those who believe that policy reforms in the water sector would be crucial for bringing about progress in human development and those who believe that economic growth itself would help solve many of the water problems, which countries are facing today (HDR, 2006, p. 66). Such debates, which are often not healthy, cause delays in deciding investment priorities in the water sector, particularly in the developing world (Biswas and Tortajada, 2001; Shah and Kumar, 2008).

The theoretical discussion on the returns on investment by countries in water infrastructure and institutions are abundant (Grey and Sadoff, 2005;

Rural Water Systems for Multiple Uses and Livelihood Security
ISBN 978-0-12-804132-1
http://dx.doi.org/10.1016/B978-0-12-804132-1.00002-0

HDR, 2006; Grey and Sadoff, 2007). The evidence available internationally to the effect that water security can catalyze human development and growth is quite robust (see World Bank, 2004a, 2006a,b; Briscoe, 2005). However, the number of regions for which these are available is not large enough for evolving a global consensus on this complex issue. Till recently, there was no comprehensive database on various factors influencing water security for a sufficient number of countries which are at different stages of the human development path and economic growth. This contributed to the complexity of the debate. The water poverty index (WPI), conceived and developed for countries by Sullivan (2002), and the international comparisons now available from a recent work by Laurence et al. (2003) for 145 countries, enable us to provide an empirical basis for enriching the debate.

The WPI is a composite index consisting of five subindices, namely, water access index, water use index, water endowment index, water environment index, and institutional capacities in the water sector (Sullivan, 2002). In order to realistically assess the water situation of a country, a new index called the sustainable water use index (SWUI) was derived from WPI (Shah and Kumar, 2008; Kumar, 2009). In this chapter, we provide empirical analyses using a global database on SWUI and many other water and development indicators, such as the global hunger index (GHI), human poverty index (HPI), and income inequality index to enrich the debate "how water security is linked to human development, poverty reduction and inclusive growth."

2.2 OBJECTIVES, HYPOTHESIS, METHODS, AND DATA SOURCES

In this chapter, we analyze the following: (1) the nature of linkage between the water situation of a country comprising improved water access and use, water environment and institutional capacities in the water sector, and its economic growth; (2) the nature of linkage between water situation and income inequality and poverty; and, (3) the role of large water storages in reducing hunger and boosting economic growth of countries which fall in hot and arid, tropical climates. But, before we do that, we first present the global debate on the link between water security, human development, and economic growth.

We have three propositions. First: improving the water situation through investments in water infrastructure, institutions and policies would help ensure economic growth through the human development route. Second: nations can achieve reasonable progress in human development even at low levels of economic growth, through investment in water infrastructure, and welfare policies. Third: countries need to invest in building large water storages to support economic prosperity, and ensure water security for social advancements. The hypotheses are: (1) improved water situation supports economic growth through the human development route; and (2) countries, which are in tropical climates with aridity, can support their economic growth through enhancing per capita reservoir storage that improves their water security.

The values of the sustainable water use index were calculated by adding up the values of four of the subindices of the water poverty index, namely, water access index, water use index, water environment index, and water capacity index. All the subindices have values ranging from 0 to 20. The maximum value of SWUI for a country therefore is 80.

The first hypothesis was tested using a regression of global data on: sustainable water use index (SWUI) and per capita GDP (purchasing power parity (ppp) adjusted); SWUI and GHI; and SWUI and HDI. GHI is an indicator of the proportion of the population living in undernourished conditions and the child mortality rate (see Doris, 2006).

Since regression between SWUI and HDI showed a strong relationship ($R^2 = 0.80$), the causality, ie, whether SWUI influences GDP growth or vice versa, can be tested by running regression between the per capita GDP and a decomposed HDI, which contain the indices for health and education. Alternatively, it was tested by running a two-stage least-square method, with SWUI as the predictor variable, HDI as the instrumental variable, and per capita GDP as the dependent variable. The underlying premise is that if economic growth drives the water situation, then it should change the indicators of human development that are independent of income levels, such as health and education, and those which are interrelated with the water situation. The second hypothesis was tested by analyzing the link between per capita GDP (ppp adjusted) and per capita dam storage (m^3/annum) of 22 selected countries falling in hot and arid tropical climate.

Data on per capita GDP and HDI were obtained from Human Development Report (2009). Data on GHI for 117 countries were drawn from Doris (2006). Data on WPI for 145 countries were obtained from Laurence and Sullivan (2003). Data on dam storage and human population in 24 countries were obtained from FAO AQUASTAT-2006. Data on income inequality (Gini coefficients) for 125 countries were obtained from the Human Development Report (2009) of UNDP. Data on human poverty index (HPI) for 113 countries, for the year 2007 were obtained from the Human Development Report (2009) of UNDP.

We also illustrate how rural water systems, which meet multiple needs, namely, drinking and cooking, domestic uses, irrigation, watering of vegetable gardens, livestock production, fisheries, brick-making, can help improve human health, livelihoods, and economic conditions of rural people, and are important for achieving all-round water security, by providing unique models of multiple-use water systems around the developing world.

2.3 THE GLOBAL DEBATE ON WATER, DEVELOPMENT, AND GROWTH

The debate on the linkage between water, growth, and development is mounting internationally. While the general view of international scholars, who support large water resource projects, is that increased investment in water projects such as irrigation, hydropower, and water supply and sanitation acts as an engine of growth in the economy, while supporting progress in human development (for instance, see Braga et al., 1998; Briscoe, 2005; HDR, 2006), they harp on the need for investment in water infrastructure and institutions. Grey and Sadoff (2007) suggest that there is a minimum platform of water security, achieved through the right combination of investment in water infrastructure and institutions and governance, which is essential if poor countries are to use water resources effectively and efficiently to achieve rapid economic growth to benefit vast numbers of their population. They suggest an S-curve for growth impacts of investment in water infrastructure and institutions in which returns continue to be nil for early investments. They argue that for poor countries, which experience highly variable climates, the level of investment required to reach the tipping point of water security[1] would be much higher as compared to countries that fall in temperate climate with low variability. But, they suggest that for developing countries, the returns on investment in infrastructure would be higher than in management, and vice versa in the case of developed countries.

Many environmental groups, on the other hand, advocate small water projects which, according to them, can be managed by the communities. The solutions advocated are: watershed management; small water-harvesting interventions; community-based water supply systems; and, microhydroelectric projects (D'Souza, 2002; Dharmadhikary, 2005).

[1] Beyond which the investment in water infrastructure and institutions yields positive growth impacts.

The proponents of sustainable development paradigms believe that the ability of a country to sustain its economic growth depends on the extent to which natural resources, including water, are put to efficient use through technologies and institutions, thereby reducing the stresses on environmental resources (Drexhage and Murphy, 2010). Here, the focus is on initiating institutional and policy reforms in the water sector. An alternative view suggests that countries would be able to tackle their water scarcity and other problems relating to the water environment at advanced stages of economic development (Shah and Koppen, 2006). They argue that standard approaches to water management in terms of policies and institutions work when water economies become formal, which are found at an advanced stage of economic development of nations.

International literature provides many clues to the fact that water security has the potential to promote inclusive growth. For instance, access to safe water and sanitation can partly determine income. The marginal productivity per unit of water, measured in terms of good health, longevity, or income is much greater for the poor than for the rich (World Water Council, 2000; van Koppen et al., 2009; Jha, 2010). Secured water for irrigation, while enabling the poor landholding communities to grow food for their own consumption (Kumar, 2003), would generate sufficient employment in rural areas and lower cereal prices (Perry, 2001) and supply cereals to the markets. In poor and developing countries, where a large section of the population depends on agriculture and rural wage labor, this is likely to have significant distributional effects on income.

2.4 WATER AND INCLUSIVE GROWTH

Before we begin to answer this complex question of "what drives what," we need to understand what realistically represents the water richness or water poverty of a country. A recent work by the Kellee Institute of Hydrology and Ecology which came out with international comparisons on water poverty of nations had used five indices, namely, water resources endowment; water access; water use; capacity-building in the water sector; and water environment, to develop a composite index of water poverty (see Laurence and Sullivan, 2003).

Among these five indices, we chose four indices as important determinants of the water situation of a country, and the only subindex we excluded was the water resources endowment. We consider that this subindex is more or less redundant, as three other subindices, namely, water access, water use, and water environment take care of what the resource endowment is expected to provide. Our contention is that natural water resource endowment becomes an important determinant of the water situation of a country only when governance is poor and institutions are ineffective, adversely affecting the community's access to and use of water, and water environment. Examples are the droughts in sub-Saharan African countries. This argument is validated by a recent analysis which showed a strong correlation between rainfall failure and economic growth performance in these countries. That said, all the four subindices we chose had significant implications for socioeconomic conditions, and are influenced by institutional and policy environment, and therefore have a human element to them. Hence, such a parameter will be appropriate to analyze the effect of institutional interventions in the water sector on the economy.

It is being hypothesized that the overall water situation of a country (or SWUI) has a strong influence on its economic growth performance. This is somewhat different from the hypothesis postulated by Shah and Koppen (2006), wherein they have argued that economic growth (GDP per capita) and HDI are determinants of water access poverty and the water environment.

It is important to provide empirical evidence to this. Worldwide, experiences show that improved water situation (in terms of its access to water; levels of use of water; the overall health of water environment; and enhancing the technological and institutional capacities to deal with sectoral challenges) leads to better human health and environmental sanitation; food security and nutrition; livelihoods; and greater access to education for the poor (HDR, 2006). This aggregate impact can be segregated with irrigation having a direct impact on rural poverty (Bhattarai and Narayanamoorthy, 2003; Hussain and Hanjra, 2003); irrigation having an impact on food security, livelihoods, and nutrition (Hussain and Hanjra, 2003), with positive effects on productive workforce; and domestic water security having positive connotations for health, environmental sanitation, with spin-off effects on livelihoods and nutrition (positive), school dropout rates (negative), and productive workforce.

According to the Human Development Report (2006), only one in every five people in the developing world has access to an improved water source. Dirty water and poor sanitation account for the vast majority of the 1.8 million child deaths each year (almost 5000 every day) from diarrhea—making it the second-largest cause of child mortality. In many of the poorest countries, only 25% of the poorest households have access to piped water in their homes, compared with 85% of the richest. Diseases and productivity losses linked to water and sanitation in developing countries amount to 2% of GDP, rising to 5% in sub-Saharan Africa—more than the aid the region gets. Women bear the brunt of responsibility for collecting water, often spending up to 4h a day walking, waiting in queues, and carrying water; water insecurity linked to climate change threatens to increase malnutrition to 75–125 million people by 2080, with staple food production in many sub-Saharan African countries falling by more than 25%.

The strong inverse relationship between SWUI and the global hunger index (GHI), developed by IFPRI for 117 countries, provides broader empirical support for some of the phenomena discussed above. In addition to these 117 countries for which data on GHI are available, we have included 18 developed countries. For these countries, we have considered zero values, assuming that these countries do not face problems of hunger. The estimated R^2 value for the regression between SWUI and GHI is 0.60. The coefficient is also significant at the 1% level. It shows that with an improved water situation, the incidence of infant mortality (below 5 years of age) and impoverishment reduces. In that case, an improved water situation should improve the value of human development index, which captures three key spheres of human development: health, education, and income status.

Thus, all the subindices of HDI have strong potential to trigger growth in a country's economy, be it educational status; life expectancy; or income levels. When all these factors improve, they would have a synergetic effect on the economic growth. The growth, which occurs from human development, would also be "broad-based" and inclusive. Hence, the "causality" of water as a prime driver for economic growth can be tested if we are able to establish a correlation between water situation and HDI. This we will examine at a later stage.

Before that, let us first look at how the water situation and economic growth of nations are correlated. Regression between sustainable water use index (SWUI) and per capita GDP (ppp adjusted) for the set of 145 countries shows that it explains the level of economic development to an extent of 69% (Fig. 2.1). The coefficient is significant at the 1% level. The relationship between SWUI and per capita GDP is a power function. Any improvement in the water situation beyond a level of 50 in SWUI leads to exponential growth in per capita GDP.

This only means that for countries to be on the track of sustainable growth, the following steps are required: (1) investment in infrastructure, and institutional mechanisms and policies to: (a) improve access to water for

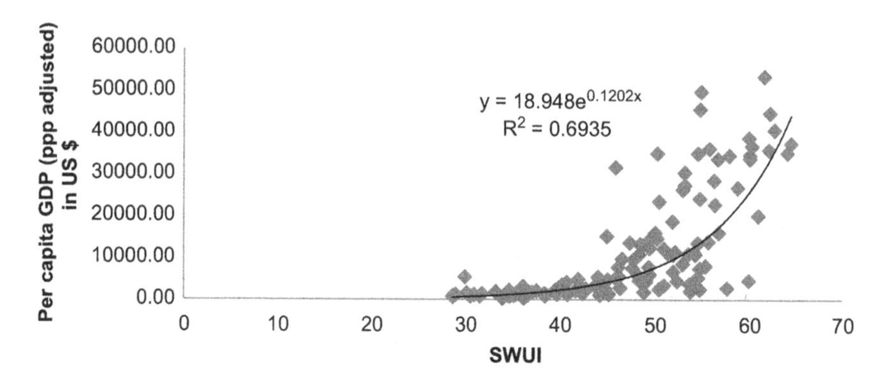

Figure 2.1 Per capita GDP (ppp adjusted) versus SWUI.

all sectors of use and across the board, (b) enhance the overall level of use of water in different sectors, and (c) regulate the use of water, reduce pollution, and provide water for ecological services; and (2) investment in building human resources and technological capabilities in the water sector to tackle new challenges in the sector. Regression with different indices of water poverty against economic growth levels shows that the relationship is less strong, meaning all aspects (water access, water use, water environment, and water sector capacity) are equally important to ensure growth.

Major variations in economic conditions of countries having the same levels of SWU can be explained by the economic policies which the country pursues. Some countries of central Asia, namely, Uzbekistan, Kyrgyzstan, and Turkmenistan; and Latin American countries, namely, Ecuador, Uruguay, Colombia, and Chile have values of SWUI as high as North America and northern European countries but are at much lower levels of per capita GDP. While North America and north, west and southern European countries have capitalist and liberal economic policies, these countries of old Soviet Bloc and Latin America have socialist and welfare-oriented policies.

2.4.1 Can Water Security Ensure Economic Growth?

International development discussions are often characterized by polarized contentions on whether money or policy reform is more crucial for progress in human development (various authors as cited in HDR, 2006, p. 66). Scholars have already discussed the two possible causal chains, one that runs between economic growth and human development, and the other that runs between human development and economic growth (Ranis, 2004). The causality in the first case occurs when resources from national income are allocated to activities that contribute to human development. Ranis (2004) argued that a low level of economic development would result in a vicious cycle of low levels of human development and a high level of economic development would result in the virtuous cycle of high levels of human development. Whereas in the second case, as indicated by several evidences, better health and nutrition lead to better productivity of the labor force (Behrman, 1993; Cornia and Stewart, 1995). Education opens up new economic opportunities in agriculture (Schultz, 1975; Rosenzweig, 1995), impacts on the nature and growth of exports (Wood, 1994), and results in greater income equality, which in itself results in economic growth (Bourguignon and Morrison, 1990; Psacharopolous et al., 1992; Bourguignon, 1995; Ranis, 2004).

If the stage of economic development determines a country's water situation rather than the reverse, the variation of human development index, should be explained by variation in per capita GDP, rather than water situation in orders of magnitude. We have used data for 145 countries to examine this closely. The regression shows that per capita GDP explains HDI variations to an extent of 90%. The regression equation was $Y=0.129\ln(X)-0.398$. But, it is important to remember that HDI already includes per capita income, as one of the subindices.

Therefore, analysis was carried out using decomposed values of HDI, after subtracting the per capita income index, the graphical representation of which is presented in Fig. 2.2. The regression value came down to 0.75 ($R^2=0.75$) when the decomposed index, which comprises education index and life expectancy index, was run against per capita GDP. What is more striking is the fact that 21 countries having per capita income below 2000 dollars per annum have medium levels of decomposed index. Again 50 countries having per capita GDP (ppp adjusted) less than 5000 dollars per

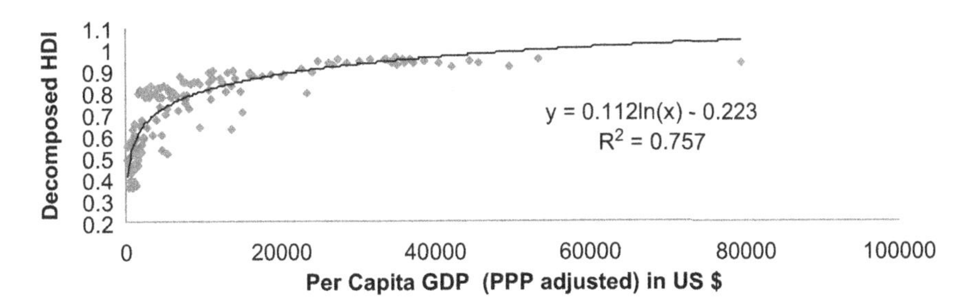

Figure 2.2 Decomposed HDI versus per capita GDP (ppp adjusted) for 2007.

annum have medium levels of decomposed HDI. Significant improvements in HDI values (0.30–0.9) occur within the small range in per capita GDP. The remarkable improvement in HDI values with minor improvements in economic conditions, and then "plateauing" means that improvement in HDI is determined more by factors other than economic growth. Our contention is that the remarkable variation in HDI of countries belonging to the low-income group can be explained by the quality of governance in these countries, ie, whether good or poor.

Many countries that show high HDI also have good governance systems and practices, and institutional structures to ensure good literacy and public health. For instance, Hungary in eastern Europe; some countries of Latin America, namely, Uruguay, Guatemala, Paraguay, Nicaragua, and Bolivia; and countries of the erstwhile Soviet Union, namely, Turkmenistan, Kyrgyzstan, and Armenia have welfare-oriented policies. They make substantial investments in water, health, and educational infrastructure.[2]

Incidentally, many countries, which have extremely low HDI, have highly volatile political systems, and ineffective governance and corruption. The investments in building and maintenance of water infrastructure are consequently very poor in these countries (Shah and Kumar, 2008) in spite of huge external aid. Sub-Saharan African countries, namely, Angola, Benin, Chad, Eritrea, Ethiopia, Burundi, Niger, Togo, Zambia, and Zimbabwe; and Yemen in the Middle East belong to this category. Sub-Saharan Africa has the lowest irrigated to rain-fed area ratio of less than 3% (FAO, 2006, Figure 5.2, p. 177), whereas Ethiopia has the lowest water storage of 20 m^3/capita in dams (World Bank, 2005). How water security decoupled human development and economic growth in many regions of the world was illustrated in the human development report (HDR, 2006, pp. 30–31).

The public expenditure on health and education is extremely low in these African countries and Yemen when compared to the many other countries which fall under the same economic category (below US $ 5000 per capita per annum). Over and above, the pattern of public spending is more skewed toward the military (source: HDR, 2006, Table 19, pp. 348–351). Besides, access to water supply and sanitation is much higher in the countries which have higher HDI, as compared to those countries which have very low HDI (based on data in HDR, 2006, Table 7, pp. 306–309).

[2] For instance, the USSR had invested in a major way for building hydraulic infrastructure in central Asia (HDR, 2006). As a result, they attain high HDI even at a low level of economic growth.

Some of the striking features of these regions are the high incidence of water-related diseases such as malaria and diarrhea, high infant mortality, and high school dropout rate mainly due to lack of access to safe drinking water; and scarcity of irrigation water in rural areas, poor agricultural growth, high food insecurity and malnutrition (source: based on HDR, 2006). Consequently, their HDI is very low.

2.4.2 Linking Human Development With Water Security

Contrary to the above-discussed scenario, regression between sustainable water use index and HDI shows that it explains variation in HDI in a much better way than the level of economic development. This is in spite of the fact that the human development index as such does not include any variable that explicitly represents access to and use of water for various uses; overall health of water ecosystem; and capacities in the water sector as one of its subindices. The R^2 value was 0.80 against 0.75 in the earlier case when per capita GDP is run against decomposed HDI (Fig. 2.3). Also, the coefficient was significant at the 1% level. It means that variation in human development index can better be explained by *water situation* in a country, expressed in terms of sustainable water use index, than the ppp adjusted per capita GDP (Kumar, 2009).

Now, such a strong linear relationship between sustainable water use index and HDI explains the exponential relationship between sustainable water use index and per capita GDP, discussed in the earlier section, as the improvements in subindices of HDI contribute to economic growth in its own way (ie, per capita GDP $= f\{EI, HI\}$; here EI is the education index, and HI is the health index).

This is further reinforced by the empirical relationship between per capita GDP (ppp adjusted) and HDI (for the year 2007) of nations, which

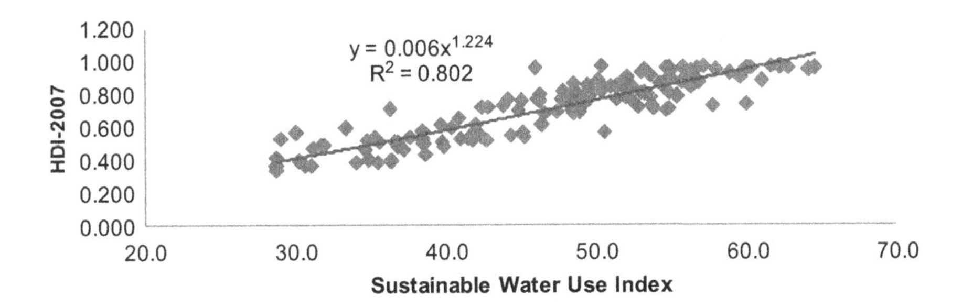

Figure 2.3 HDI-2007 versus SWUI.

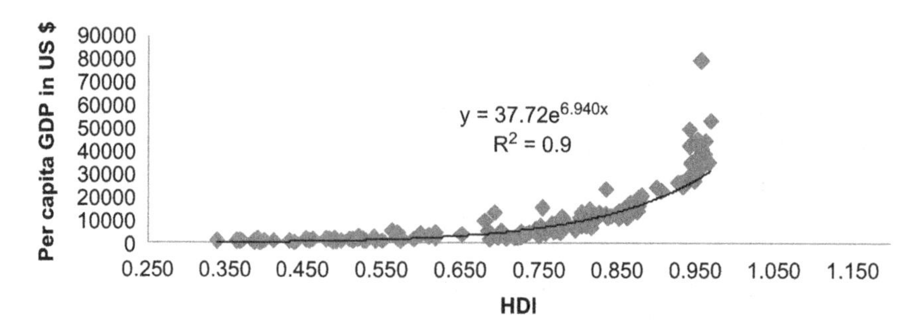

Figure 2.4 Per capita GDP (ppp adjusted) versus HDI-2007.

show that per capita GDP increases exponentially with increase in HDI (Fig. 2.4). Higher levels of HDI result in much higher levels of income. As the graph shows, there are some countries, which have medium level of development, but shows very high income. South Africa is one example. They fall off the trajectory which the majority of countries follow. It is important to remember here that this country suffers from high levels of income inequity.

While an alternative to analyze the impact of a country's water situation on its economic growth performance is to look at the historical data on cumulative investments in water sector, water access and use by population in different sectors, change in water environment, and economic conditions for individual nations. Time series data on these are seldom available on a time series basis. Under such a circumstance, the best way to go ahead is to analyze the impact of natural water endowment, ie, rainfall on economic growth in a situation where investments in infrastructure and institutions and governance mechanisms for improving water access, water use, and water environment are poor. The reason is that under such situations, the water access, water use, and water environment would be highly dependent on natural water endowment.

There cannot be a better region than sub-Saharan Africa to illustrate such effects. An analysis carried out in 2004 showed a strong correlation between rainfall trend since 1960s and GDP growth rates in the region during the same period, which argued that the low economic growth performance could be attributed to a long-term decline in rainfall which the region experienced (Barrios et al., 2004). Such a dramatic outcome of rainfall failure can be explained partly by the failure of the governments to build sufficient water infrastructure. Sub-Saharan Africa has the smallest proportion of its cultivated area (<3%) under irrigation (HDR, 2006). Due to this reason, reduction in rainfall leads to a decline in agricultural production, food insecurity, malnutrition, loss of employment opportunities, and an overall drop in economic growth in rural areas.

In the absence of empirical evidence which can establish the nature of the link between SWUI and economic growth condition, we further test the causality of SWUI acting as a determinant of economic growth by running a two-stage least-square method with HDI as the instrumental variable, SWI as the predictor variable, and per capita GDP (ppp adjusted) as the dependent variable. Here, it is assumed that SWUI will have a major bearing on HDI. The link has also been established through empirical analysis. The results are presented below. It shows an R^2 value of 0.50, at the 1% level of significance. The beta coefficient is 1192 and the constant is −44,636.0.

The foregoing analyses suggest that improving sustainable water use index, which is reflective of how good the water situation of a country, is of paramount importance if we need to achieve sustained growth. It would be rather improper to assume that a country can wait till its economy improves to a certain level to start tackling its water problems. While the natural water endowment in both qualitative and quantitative terms cannot be improved through ordinary measures, the *water situation* can be improved through economically efficient, just and ecologically sound development and use of water in river basins.

2.4.3 Linking Water Security With Inclusive Growth

We have already tested the causality of water security as a driver of economic growth by establishing the strong linkage between water security and human development. The relationship was found to be linear with an R^2 value of 0.80. Scholars have earlier shown the negative impact of education on income inequality based on empirical evidence from selected countries around the world (source: based on Bourguignon and Morrison, 1990; Psacharopolous et al., 1992; Bourguignon, 1995; Ramirez et al., 1997). But, now with the availability of data on income inequality for a large number of countries (for the year 2007) from the Human Development Report of 2009, the impact of improved water security on inclusiveness of economic growth can be empirically tested by analyzing the link between HDI and income inequity existing in the countries. The data on income inequality, expressed in terms of Gini coefficient of income, was obtained for 125 countries from the Human Development Report of 2009 (HDR, 2009). Higher values of Gini coefficient indicate greater income inequality among people of the nation considered (UNDP, 2009).

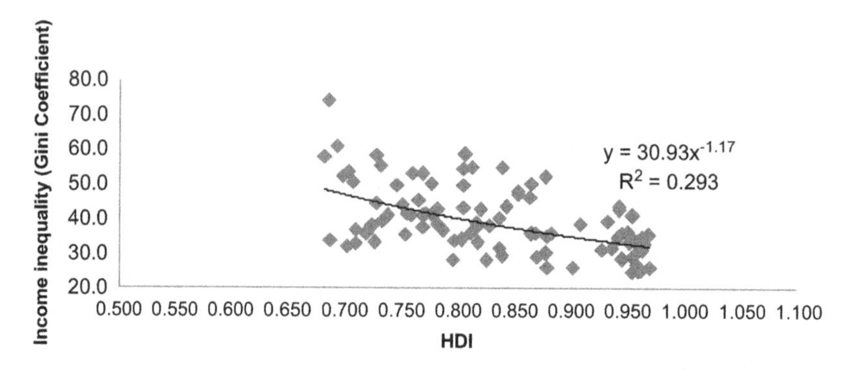

Figure 2.5 Income inequality versus HDI for high HDI countries.

Analysis shows that HDI has a direct positive impact on income equality, when countries with high levels of human development (above 0.65 and up to 0.971) are considered for the analysis. The R^2 value was 0.29, with an inverse exponential relationship between HDI and income inequality (Fig. 2.5). But, such a relationship did not emerge when the analysis was performed after considering data from all the low-HDI as well as high-HDI countries. The reason is some of the low-HDI countries have low-income inequality.

However, we have already seen that countries with low HDI have very low per capita GDP, though the opposite was not true. It is important to remember that having low-income inequality alone cannot be treated as a great virtue for a country, when the income levels are very low. Income equality does not have much relevance from a developmental perspective, when the average income levels are too low.

Obviously, the most desirable situation is a high average income and low-income inequality. The foregoing analysis means that when the human development of a country crosses a particular threshold level, the national wealth gets better distributed. Since, water security influences HDI positively, the type of relationship that will emerge between water security and income inequality is likely to be same as the relationship found between HDI and income inequality. This means, improving the water security of a nation would be a necessary and sufficient condition for achieving high levels of development and inclusive growth. The analysis carried out using SWI for 79 countries (for which the values are above 45), and the values of income inequality (measured as Gini coefficient) shows an inverse correlation (Fig. 2.6) with an R^2 value of 0.22, meaning variation in sustainable water use index explains variations in income inequality by 22%.

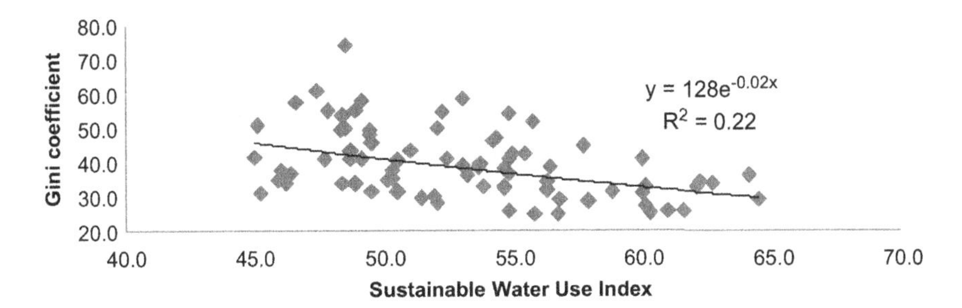

Figure 2.6 Income inequality (2007) versus sustainable water use index (2003) for selected countries.

The relationship was rather weak, when regression was run for all the 125 countries for which data on both SWI and income inequality are available. This leads to the point that the distributional effect of national income gets affected when water security and therefore human development crosses a certain threshold.

Poverty reduction is one of the objectives of economic growth policies, and economic growth is essential for poverty reduction (Pearce and Warford, 1993). We have seen that improved water security results in better economic growth conditions, through the human development path. We have also seen that there is a greater distribution of income when the water security situation exceeds certain thresholds. The next step in the line of investigation is to find out whether this has a real impact on poverty reduction. The human poverty index is a sound indicator for a cross-country comparison of the poverty situation existing in nations.[3] Analysis was carried out using the data of HPI (Human Poverty Index-1 [available for 113 countries only]) and SWUI.[4] The analysis shows a strong inverse correlation between the two (Fig. 2.7). The countries having higher values of sustainable water use index have a lower incidence of poverty. The R^2 value was estimated to be 0.68 for linear regression, which provided best fit.

[3] It is a composite measure of the number of people living below the poverty mark (expressed in terms of per cent population below $1.25 mark, per cent population below $2 mark; and per cent population living below national poverty line); adult illiteracy rate in the age group of 15–65; probability of not surviving till the age of 40; per cent population not using an improved water source; and underweight children below the age of 5 (UNDP, 2009).

[4] A different measure of Human Poverty Index (HPI-2) is followed for a group of 30 countries, most of which (25) have a very high human development index, and the rest falling under the high human development index category.

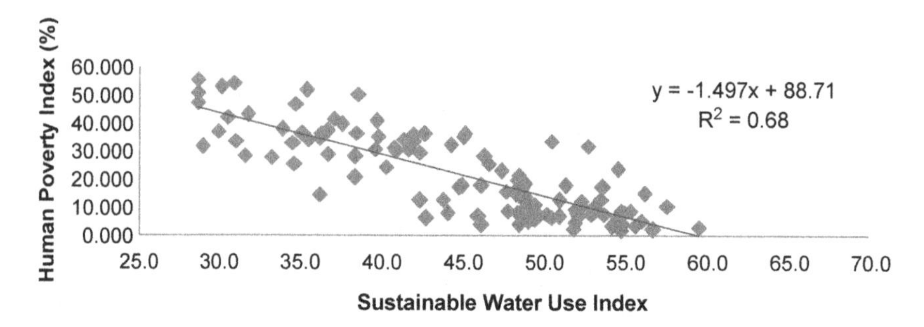

Figure 2.7 Human Poverty Index-1 (2007) versus SWUI (2003) for 113 countries.

2.5 IMPACT OF STORAGE DEVELOPMENT ON ECONOMIC GROWTH IN ARID TROPICS

Development of reservoir storage has an important role in improving the access to and use of water, the two prerequisites for improving the water situation of a region, through intensive water development in river basins might cause environmental water stress ,reducing the values of the water environmental index (Shah and Kumar, 2008; Kumar, 2009). This is evident from the direct logarithmic relationship between storage development and water security (expressed in terms of sustainable water use index) (Fig. 2.8). The R^2 value was estimated to be 0.39. Fig. 2.8 shows that countries with higher per capita reservoir storage (expressed in m^3 per capita per annum) have higher values of sustainable water use index. Major improvements in water security occurred within the range of 0–1500 m^3 per capita per annum, and leveling off thereafter.

However, the amount of storage that needs to be created to improve access to and use of water depends on the type of climatic conditions. It is also important to note that access to arable land would also be an important factor determining the storage requirements, as water requirement for agriculture would change with this. In temperate and cold climates, the demand of irrigation, the largest water use sector, would be considerably smaller compared to tropical and hot climates. Hence, the storage requirements in such regions would be much lower, and would be mainly limited to that for meeting domestic/municipal water needs and water for manufacturing. Therefore, it is logical to explore links between storage development for meeting various human needs and economic growth only in tropical and hot climates.

But, as indicated in the Human Development Report of 2006, the sheer scale of water infrastructure in rich countries is not widely recognized and

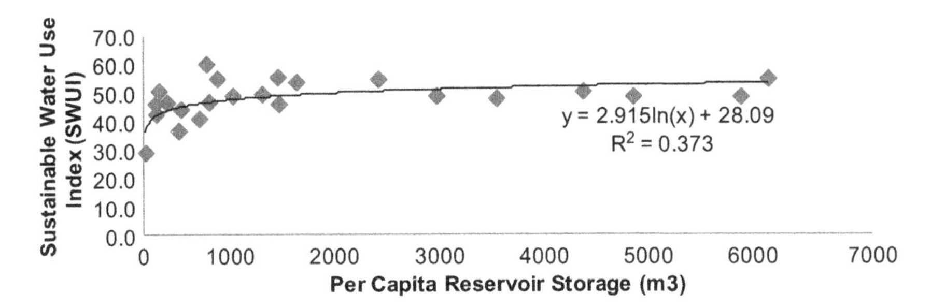

Figure 2.8 SWUI versus per capita reservoir storage.

appreciated (HDR, 2006, p.155). Many developed countries of the world that experience tropical climates had high water storage in per capita terms. The US, for instance, had created a per capita storage capacity of nearly 6000 m^3. In Australia, the 447 large dams alone create a total storage of 79,000 million cubic metres (MCM) per annum, providing per capita water storage of nearly 3808 m^3 per annum. Aquifers supply another 4000 MCM per annum. China, the fastest growing economy in the world, has a per capita reservoir storage capacity of 2000 m^3 per annum through dams, and an actual storage of nearly 360 m^3 per capita (Kumar, 2009).

When compared to these figures, India, which is still developing, has a per capita storage of only 220 m^3 per annum. Though a much higher level of withdrawal of nearly 600 m^3 per capita per annum is maintained by the country, a large percentage of this (231 BCM per annum or nearly 217 m^3 per capita per annum) comes from groundwater draft and there is increasing evidence to suggest that this won't be sustainable.[5] Ethiopia, the poorest country in the world, has a per capita storage of 20 m^3 per annum. These facts also strengthen the argument that economic prosperity that a country can achieve is a function of available per capita water storage.

Regression analysis of per capita water storage and the per capita GDP (ppp adjusted) for a group of 24 countries falling in the arid and semiarid tropics shows a strong relationship between the level of storage development and a country's economic prosperity (Fig. 2.9). The R^2 value is 0.55, and the coefficient is significant at the 1% level. The strong relationship can be explained in the following way. Storage reservoirs reduce risks and improve water security. In many regions, investments in hydraulic infrastructure

[5] As discussed in a recent work by Kumar (2007), many semiarid areas are already facing problems of groundwater overdraft, with serious socioeconomic and ecological consequences.

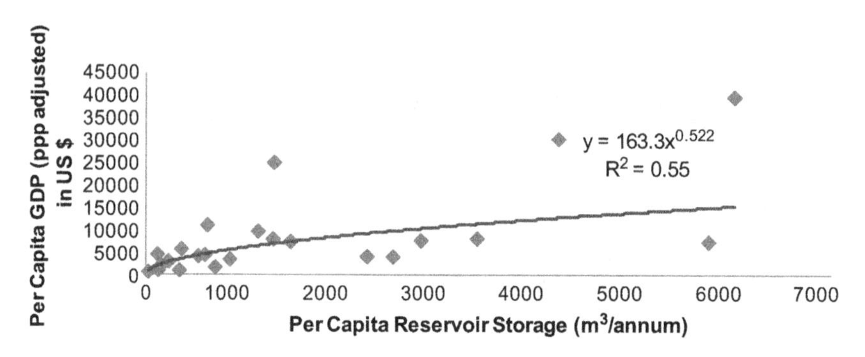

Figure 2.9 Per capita GDP versus per capita reservoir storage (24 countries).

supported economic prosperity and social progress, though in some regions it caused environmental damage (Grey and Sadoff, 2005; HDR, 2006; based on various authors, p. 140).[6]

The returns on investments in building water storages were quite visible in India. The analysis using panel data on gross irrigated area and rural poverty rate for 14 states showed the poverty-reducing effect of irrigation, with the lowest rate of poverty found in Punjab which had the highest level of gross irrigated area, which reduced over time from 1973–74 to 1993–94 (Bhattarai and Narayanamoorthy, 2003). The Bhakra-Nangal Project transformed the economy of Punjab. The almost perennial water supply from the project enabled farmers in this region to intensify cultivation with irrigated paddy and wheat, making it the country's breadbasket.[7] In Gujarat state of western India, the impact of the yet-to-be-completed Sardar Sarovar Project (SSP) in reviving the agricultural production, after it experienced a major dip following two consecutive years of drought (1999 and 2000), has been remarkable (Kumar et al., 2010). The project, which brings water from

[6] Since 1920, the US Army Corps of Engineers had invested a sum of $200 billion on flood management and mitigation alone, yielding a benefit of $700 billion. The Tennessee Valley Authority, which built dams for hydropower, transformed a flood-prone, impoverished region in the United States, with some of the worst human development indicators, into an agriculturally prosperous region. In Japan, heavy postwar investments in infrastructure supported rapid development of hydropower, flood control, and irrigated agriculture. The returns from these investments were tremendous. Until World Water II, the floods and typhoons had resulted in losses often amounting to 20% of GNI, whereas since the 1970s, the losses never exceeded 1% of the GNI (HDR, 2006, p. 156).

[7] Now, 90% of the cropped area in Punjab is irrigated, with paddy and wheat accounting for three-fourths of it. Despite having only less than 2% of the geographical area of the country, Punjab accounts for 10% of rice production and 20% of wheat production in India. Agriculture accounts for 40% of the state's GDP, in the state which has the highest per capita GDP amongst all Indian states (Cummings et al., 2006).

the water-rich south Gujarat, to the water-scarce regions of north Gujarat, Saurashtra and Kachchh, reduces the imbalances in water availability and demand in different regions of the state.

The potential positive impact of water infrastructure on economic growth in regions that experience seasonal climates, rainfall variability, and floods and droughts can be better demonstrated by the economic losses that water-related natural disasters cause in the regions which lack them (Kumar, 2009). For instance, in Ethiopia, deviation in per capita GDP from the normal values during the 20-year period from 1980 to 2000 correlated with departure of annual rainfall from normal values (World Bank, 2006a). In Kenya, economic losses due to floods during 1997–98 were to the tune of 11% of the national GDP, whereas that due to droughts during 1998–2000 was 16% of the GDP (World Bank, 2004b, 2006b). In the Indian state of Gujarat, the value of agricultural output dropped from Rs. 268.37 billion in 1998–99 to Rs. 189.0 billion in 2000–01 following the droughts in 1999 and 2000 (Kumar et al., 2010, Figure 1, p. 4).

Nevertheless, the overall economic growth impact of water storage depends on the nature of uses for which the resources are developed, the effectiveness of the institutions that are created to allocate the resource and the nature of institutional and policy regimes that govern the use of the resource. As we have seen in the case of incidence of hunger in Zambia and Zimbabwe, use of water storages for hydropower generation did not help to improve the overall economic condition of the people (Kumar, 2009). Though the per capita water storage in Israel is quite low (nearly 150 m^3 per annum), the efficiency with which water is used in different sectors is extremely high. Nearly 90% of the country's irrigated area is under micro-irrigation systems. A large portion of the water used in urban areas is recycled and put back to use for irrigation. Water is not only priced on volumetric basis, water allocation for irrigation is also rationed (Kumar, 2009).

One could also argue that access to water could be better improved through local water resources development interventions, including small water-harvesting structures or through groundwater development. As a matter of fact, environmental activists advocate decentralized small water-harvesting systems as alternatives to large dams (see Agarwal and Narain, 1997). Small water-harvesting systems have been suggested for water-scarce regions of India (Agarwal and Narain, 1997; Athavale, 2003), and the poor countries of sub-Saharan Africa (Rockström et al., 2002). However, evidence suggests that they cannot make any significant dent in increasing water supplies in countries like India due to the unique hydrological regimes, and can also prove to be prohibitively expensive in many situations (Kumar et al., 2006, 2008). Also, to meet large concentrated demands in urban and industrial areas, several thousands of small water-harvesting systems would be required. The type of engineering interventions[8] and the economic viability of doing the same are open to question. Recent evidence also suggests that small reservoirs get silted up much faster than large ones (Vora, 1994), a problem for which large dams are criticized for the world over (see McCully, 1996).

As regards groundwater development, intensive use of groundwater resources for agricultural production is proving to be catastrophic in many semiarid and arid regions of the world, including some developed countries like Spain, Mexico, Israel, Australia, and parts of the US (Kumar, 2007), and developing countries such as India, China, Pakistan, Yemen, and Jordan (HDR, 2006). However, some of the developed countries have achieved some degree of success in controlling it through the establishment of management regimes (Kumar, 2007) with physical and institutional interventions like in the western US, or through physical interventions alone, like in Israel, or through institutional interventions such as formal water markets, like in the Murray–Darling basin of Australia.

However, in the basins facing problems of environmental water scarcity and degradation (see Smakhtin et al., 2004) due to large water projects, river flows are appropriated and transferred for various consumptive needs. Some of these basins include the Colorado river basin in the western US; Yellow river basin in northern China; Aral sea basins, namely, Amu Darya and Syr Darya in Central Asia; Indus basin in Pakistan and India; basins of northern Spain; Nile basin in northern Africa; basins of Euphrates, Tigris; the Jordan river; Cauvery, Krishna and Pennar basins of peninsular India; river basins of western India including Sabarmati, Banas, and Narmada, located in Gujarat, Rajasthan, and Madhya Pradesh in India. Most of the water demands they meet are agricultural.[9] They are also agriculturally prosperous regions. Not only do they meet the food requirements of the region, most of these basins export a significant chunk of the food to other regions of the world, including some of the water-rich regions, within the country's territory (Yang, 2002 for China; Amarasinghe et al., 2004 for Indus basin and peninsular region in India; Kumar and Singh, 2005 for many water-scarce countries of the world).

[8] Complex engineering interventions would be required for collecting water from such number of small water harvesting and storage systems, and then transporting to a distant location in urban areas.

[9] In Murray–Darling basin, 90% of the annual flows are diverted for agricultural use.

Strikingly, wherever aquifers are available for exploitation, these regions have experienced problems of groundwater overdraft, though some developed countries developed the science to deal with it. The most glaring examples are aquifers in the western US, aquifers in the countries of the Middle East including Yemen, Iran, and Jordan; aquifers in Mexico; north China plains (Molden et al., 2001); alluvial aquifers of Indus basin areas in India; hard rock aquifers of Peninsular India; and aquifers in western and central India (GOI, 2005).

Without these large surface water projects, agricultural growth might have caused a far more serious negative impact on groundwater resources in these regions. In fact, it is this surface water availability, which to a great extent helps reduce the dependence of farmers as well as cities on groundwater (Kumar, 2009). For instance, imported water from Indus basin through canals in Indian and Pakistan Punjab sustain intensive groundwater use in the regions, through continuously providing replenishment through return flows from surface irrigation (Hira and Khera, 2000; Ahmed et al., 2004; Kumar, 2007). In India, water imported from a large reservoir, Sardar Sarovar in Narmada basin in Southern Gujarat, has started supplying water to rejuvenate the rivers in environmentally stressed basins of north Gujarat (Kumar et al., 2010).

2.6 IMPACT OF STORAGE DEVELOPMENT ON MALNUTRITION AND CHILD MORTALITY

Storage development is found to have a direct impact on malnutrition and infant mortality, the factors considered in estimating the global hunger index, when we considered zero values of GHI for developed countries, namely, the US, Australia, and Spain for which data on GHI are not available (Kumar, 2009). Regression shows an R^2 value of 0.59. The relationship between per capita storage and GHI is inverse, logarithmic. It means greater water storage reduces the chances of human hunger. This inverse relationship can be explained this way. For the sample countries, the ability to cultivate the available arable land intensively would increase with the amount of water storage facilities available. As HDR (2006, p. 174) notes "Water security in agriculture pervades all aspects of human development." Increased availability of irrigation water reduces the risk of crop failure, enhances the ability of farmers to produce more crops to improve their own domestic consumption of food, and takes care of the cash needs. Also, increased irrigated production improves the food and nutritional security of the population at large by lowering cereal prices in the region in question as the gap between cereal demand and supplies is reduced (Hussain and Hanjra, 2003 as cited in HDR, 2006, p. 175).

This was more evident in India than anywhere else, where irrigation expansion through large storages had contributed nearly 47 million tons of additional cereals to India's breadbasket (Perry, 2001, p. 104). The most illustrious example of recent times is the impact of the Sardar Sarovar Project, which is yet to be completed on food production and agricultural growth in Gujarat. The availability of surface water through canals had motivated farmers in south and central Gujarat to take up paddy and wheat farming and achieve bumper food grain production in recent years. Shah and Kumar (2008) made a rough estimate of the positive externality it created in terms of lowering food prices for the consumers in India as US $20 per ton of cereals. One could also argue that rich countries could afford to import food. However, what is important is that water played a large role for these countries in achieving a certain level of economic growth and prosperity, by virtue of which they can now afford to import food instead of resorting to domestic production. The exceptions are some of the oil-rich countries of the Middle East, which do not have an agrarian base, but are economically prosperous.

Contrary to what was found from our analysis of 22 countries, countries such as Zambia and Zimbabwe have large storages but have very high GHI (Kumar, 2009). They were not included in our analysis. These countries use their water storages for creating hydropower, which is sold to South Africa, and they earn revenue from it.[10] Hence, storage development does not lead to increased agricultural production in these countries. The GHI values are very high for these countries (Wiesmann, 2006). In such a situation, the impacts on food security would generally be seen only after many years. But in the case of these sub-Saharan African countries, three decades of droughts and rainfall reduction have significantly affected hydropower generation as well (McCully and Wong, 2004).

2.7 MULTIPLE-USE WATER SYSTEMS FOR ALL-ROUND WATER SECURITY IN RURAL AREAS

In most developing countries, economies are largely agrarian, and more people live in rural areas where agriculture and livestock rearing are the major sources of livelihood. Rural areas require water for drinking and

[10] Most of it comes from just one hydropower dam, Kariba, built in 1955–59 in the Zambezi river basin.

cooking, domestic use, livestock use, fisheries and irrigation of farms, watering of kitchen gardens, and in small reservoirs for recreation. Since developing economies face scarcity of financial resources for investment in infrastructure building, especially building of water infrastructure, building separate systems for single use (such as reservoirs with canals for irrigation, ponds for fishing, wells/hand pumps for water supply, ponds for vegetable gardens, reservoirs for recreational uses) would prove to be very expensive. In such cases, the rural water systems should be designed for multiple water needs, such as drinking and cooking, domestic use, livestock use and irrigation, rather than as single-use systems.

For instance, an expensive irrigation system, which takes water from a water-rich region to the hundreds of thousands of rural farms in a distant water-scarce region at a heavy cost, should also be able to augment the drinking water supply in the nearest villages, where the communities live, and also augment the local reservoirs. The augmenting local reservoirs can help fisheries, augment the domestic water supply, and provide recreational benefits, while augmenting groundwater recharge. The economic logic is that the returns on investment in irrigation would be generally low, and the same can be offset by significant socioeconomic, ecological, and recreational benefits that can be derived through a small additional investment for increasing the storage capacity of local reservoirs and providing a pipeline to take the water by a canal system to village sites.

An extensive review was undertaken by GSDA/IRAP (2013) on multiple-use water systems (MUWS) around the world. A total of 17 different types of MUWS, from countries as far and wide as India, Brazil, Colombia, Ethiopia, north-eastern Morocco, rural and urban Zimbabwe, Bolivia, China, hills of Nepal, northeastern Thailand, South Africa, Sri Lanka, Vietnam, and the flood plains of Bangladesh were covered in the review. From within India itself, systems from Karnataka, Maharashtra, and Indo Gangetic plains (IGP) were covered. The results were presented with regard to the physical settings in which the systems work in terms of hydrology, geology, and topography; the description of the MUWS, covering the physical and socioeconomic aspects of water supply and use; and the key features of the MUWS which result in their good performance.

The outcomes of the review suggest a wide heterogeneity in size, the basic physical features and the services they provide, displayed by MUWS. The source of water for MUWS varies from springs (Nepal hills) to roof rainwater catchment system, deep bore wells, ponds and tanks (northeastern Thailand) to a combination of roof water harvesting system (RWHS) in medium uplands and secondary reservoirs fed by canal seepage (IGP) to large artificial reservoirs initially built for hydropower (Brazil), offtakes from rivers, and primary, secondary, and tertiary canals of large schemes primarily built for irrigation, drainage, and flood control (Bac Hung Hai, Vietnam and Shahapur canal in Karnataka), reservoirs of small communal dams (Zimbabwe) and flood plains (Bangladesh). The size varies from small dams to large reservoirs and canal systems. The broad range of services provided by the systems are irrigation, power generation, flood protection, drainage, fisheries, homestead gardens, cattle drinking, domestic water supply, drinking water, brick making, recreation, environmental flows, cultural uses, and industrial and municipal water supply.

While some are designed by the agency as multiple-use systems, some have multiple-use systems by default. In the process, the technical features of some of the systems have undergone changes. In the Nepal hills, water from springs is collected in overhead tanks, and is distributed through two parallel distribution pipes, one for domestic water supply and the other for kitchen gardens. Where a single distribution pipe is provided, two taps are provided to the stand post to avoid conflicts over collection of water. Hence, different water use priorities of the communities are incorporated into the design itself (van Koppen et al., 2009).

In the case of the canal systems in the IGP, fish trenches and raised beds were constructed for integrated agriculture and fishery using seepage from canals. Fishes are raised in the ponds, while horticultural crops, vegetables, and pulses are grown on 3-m wide raised bunds. Additional runoff water harvested is used for providing supplementary irrigation to paddies. The seasonally waterlogged area is used for raising paddies and fish using nylon nets. Rainwater harvesting systems in the plateau regions of uplands irrigate the horticultural crops in their commands (Sikka, 2009; Haris, 2013).

In northeast Thailand, about 10 different sources of water were used in the rural households, namely, rainwater harvested from roofs and stored in large jars; expensive bottled water from shops; commercial tap (piped) water from outside the farm; traditional shallow wells; deep bore wells; ponds; tanks; nearby streams and canals; run-on water from nearby fields; and direct precipitation. Integrated farming is practiced using farm ponds. The drinking water was sourced from the roof catchment stored in large jars. The productive uses of water included: watering vegetables, spices; watering livestock (cows, poultry); keeping the fish tank adequately filled; irrigation of fruit trees and rice crop; raising of ducks and frogs (van Koppen et al., 2009).

In Brazil, all the major hydrographic basins have been impounded by the construction of large reservoirs. Initially built up with the objective of providing hydropower for energy supply, these artificial ecosystems were subjected to multiple uses in the course of time. Many reservoirs are used for recreation, tourism, and fisheries as well as for water supply. The reservoir functions are as follows: flood control, hydropower, navigation, water supply, irrigation, recreation, tourism, and agriculture. The management of natural or artificial ecosystems is now in a transition stage from reactive, local and sectoral to predictive, at watershed level using an integrated approach (Tundisi et al., 2005).

In the Limpopo basin of Zimbabwe, small communal dams are multipurpose structures whose uses have varying water consumption levels, water productivity, and economic values. The water from these dams is mainly used for the following: livestock drinking; irrigation; domestic; brick making; fishing; and recreation (Rusere, 2005; van Koppen et al., 2009).

In Columbia, different types of water supply sources exist in rural areas, such as public reticulation systems, community reticulation systems, wells, rainwater harvesting, water vendors, direct abstraction from streams. For drinking water supply, infrastructures such as intakes and pipelines are provided in this system. In a relatively small area, there are seven large water supply systems serving five communities. No separate or adapted infrastructure exists for productive and recreational uses. Use is made of the drinking water supply infrastructure. The net per capita domestic consumption is about 150 lpcd (litres per capita per day). Irrigation is practiced in 25% of the households, most of it used for watering small vegetable gardens. In a few cases, people irrigate larger terrains (about 0.6 ha) with crops such as beans. On the basis of cropping pattern, water consumption for irrigation was estimated at 471 L/household/day in the dry period. Water consumption in households with animals averaged 77 L/household/day. For recreational use, the daily consumption was 137 L for a swimming pool of 50 m^3 (Pérez et al., 2004; van Koppen et al., 2009).

In northern Morocco, irrigation canals are the main sources of water in rural areas. The water from irrigation canals is diverted and stored in storage tanks called *Jboub*. A full tank could provide a household with water for periods ranging from 1 week to more than 2 months. Apart from irrigation, water was used for drinking, watering livestock, small-scale brick marketing, and tree nurseries. Increased water allocations from the irrigation canals for the Jboub, especially outside the main irrigation season, make more water available for domestic and productive purposes. Higher water availability for domestic purposes, especially hygiene, is very important in the reduction of diarrheal diseases (Boelee et al., 2007).

The Bac Hung Hai irrigation scheme in Vietnam was originally built for three services: irrigation water supply, drainage, and flood control. The water supply network consisted of: intake that diverts from the river using pumping and lifting stations; primary canals; secondary canals; and tertiary canals. The water provisioning services offered by the system were: domestic water; irrigation; water for cattle; fishery; homestead garden; and industry and business. The regulating services were: environmental flows; flood protection; and sewage/drainage water. The supporting services were: transportation; habitat improvements (shade, cooling effect, material for flood protection); captured fisheries. The cultural services were: social functions linked to the infrastructure and management (Renault et al., 2013).

The review suggests that the systems which were originally designed for "single" *de facto* function as multiple-use water systems, and are capable of performing all three functions of water, ie, water as a social good (drinking and domestic uses, recreation), economic good (water for productive uses such as fish production, irrigation of farms and vegetable gardens, manufacturing and livestock production), and environmental good (water for regulating flows in rivers, recharging groundwater and flood control), depend on the environmental conditions of the area. Hence, in ideal conditions, they have higher potential to improve water security than "single-use systems."

2.8 SUMMARY, CONCLUSIONS, AND POLICY

A new index called SWUI was derived from WPI using four of its five subindices to assess the water situation of a country. Analysis was carried out using data on SWUI, GHI, HDI, per capita GDP, and per capita water storage in dams to understand the nature of linkage between the water situation of a country and its economic growth.

The analysis shows that improving the water situation can trigger economic conditions in a nation. As this occurs through the human development route, the growth would be inclusive. This strong linkage can be partly explained by the reduction in malnutrition and infant mortality, with an improvement in the water situation. Further, nations could achieve good indicators of development even at low levels of economic growth, through investment in water infrastructure and welfare-oriented policies. Many countries of the erstwhile Soviet Union, and communist countries of Latin America, which have low incomes, spend a significant portion of public funds in health and education, compared with many poor countries of sub-Saharan Africa, which spend much less for health and education and more on the military.

Analysis also shows that improving water security also promotes better distribution of national income through the human development route, as indicated by the strong inverse correlation between sustainable water use index and Gini coefficient of income of nations, and inverse correlation between HDI and Gini coefficient of income. Improved water security also reduces poverty, as indicated by the strong inverse relationship between sustainable water use index and human poverty index (1) for a group of 113 countries.

Countries which fall within tropical semiarid and arid climate can improve their economic conditions by enhancing their reservoir storage. Greater storage provides increased water security, which reduces the risks associated with droughts and floods. Such natural calamities, which cause huge economic losses, are characteristic of these countries. Nevertheless, the impact of storage could depend on the nature of uses for which the resources are developed, the effectiveness of the institutions that are created to allocate the resources, and the nature of institutional and policy regimes that govern the use of the resources.

Findings suggest that economically poor countries, which also show very poor indicators of human development, need not wait till the economic conditions improve to address water sector problems. Instead, they should start investing in building water infrastructure, create institutions and introduce policy reforms in the water sector that could lead to sustainable water use. Only this can support progress in human development, and sustain economic growth which is also inclusive. But, a prerequisite for hot and arid tropical countries is that they invest in large water resource systems to raise the per capita storage. More importantly, it is important that the water supply systems be designed as multiple-use systems, which consider the real needs of the local communities, rather than as "single-use systems," in order for these countries to achieve all-round water security fast, with optimum investments. This will help them fight hunger and poverty, malnutrition, infant mortality, and reduce the incidence of water-related disasters.

REFERENCES

Agarwal, A., Narain, S., 1997. Dying Wisdom: Rise and Fall of Traditional Water Harvesting Systems. Centre for Science and Environment, New Delhi.

Ahmad, M.U.D., Masih, I., Turral, H., 2004. Diagnostic analysis of spatial and temporal variations in crop water productivity: a field scale analysis of the rice-wheat cropping system of Punjab, Pakistan. Journal of Applied Irrigation Science 39 (1), 43–63.

Amarasinghe, U.A., Bharat, R.S., Aloysius, N., Scott, C., Smakhtin, V., de Fraiture, C., Sinha, A.K., Shukla, A.K., 2004. Spatial Variation in Water Supply and Demand across River Basins of India Research Report 83. International Water Management Institute, Colombo, Sri Lanka.

Athavale, R.N., 2003. Water Harvesting and Sustainable Supply in India. Published for Environment and Development Series, Centre for Environment Education. Rawat Publications, Jaipur and New Delhi.

Barrios, C.S., Bertinelli, L., Strobl, E., 2004. Trends in rainfall and economic growth in Africa: a neglected cause of the growth tragedy. In: Proceedings of the German Development Economics Conference, Zurich.

Behrman, J.R., 1993. The economic rationale for investing in nutrition in developing countries. World Development 21 (11), 1749–1771.

Bhattarai, M., Narayanamoorthy, A., 2003. Impact of irrigation on rural poverty in India: an aggregate panel data analysis. Water Policy 5 (5), 443–458.

Biswas, A.K., Tortajada, C., 2001. Development and large dams: a global perspective. International Journal of Water Resources Development 17 (1), 9–21.

Boelee, E., Hammou, L., van der Hoek, W., 2007. Multiple use of irrigation water for improved health in dry regions of Africa and South Asia. Irrigation and Drainage 56 (1), 43–51.

Bourguignon, F., 1995. Equity and Economic Growth: Permanent Questions and Changing Answers? In: Background paper prepared for the 1996 Human Development Report. New York, US: United Nations Development Programme.

Bourguignon, F., Morrisson, C., 1990. Income distribution, development and foreign trade: a cross-sectional analysis. European Economic Review 34 (6), 1113–1132.

Braga, B., Rocha, O., Tundisi, J., 1998. Dams and the environment: the Brazilian experience. International Journal of Water Resources Development 14 (2), 127–140.

Briscoe, J., 2005. India's water economy: bracing up for a turbulent future. In: Key Note Address at the 4th Annual Partners' Meet of IWMI-Tata Water Policy Research Program. Institute of Rural Management Anand, Gujarat, India.

Cornia, G.A., Stewart, F., 1995. Food: two errors of targeting. In: Stewart, F. (Ed.), Adjustment and Poverty: Options and Choices. Routledge, London, pp. 82–107.

Cummings, R., Rashid, S., Gulati, A., 2006. Grain price stabilization experiences in Asia: what have we learned? Food Policy 31 (4), 302–312.

D'Souza, D., 2002. Narmada Dammed: An Enquiry into the Politics of Development. Penguin Books India, New Delhi.

Dharmadhikary, S., 2005. Unraveling Bhakra: Assessing the Temple of Resurgent India. Manthan Adhyayan Kendra, New Delhi.

Doris, W., 2006. Global Hunger Index: A Basis for Cross Country Comparison. International Food Policy Research Institute, Washington, DC.

Druxehage, John, Murphy, D., 2010. Sustainable Development: from Brundtland to Rio 2012, background paper prepared for consideration by the High Level Panel on Global Sustainability at its first meeting. 19 September 2010. United Nations Headquarters, New York.

Food and Agriculture Organization of the United Nations (FAO), 2006. The AQUASTAT Database. Food and Agriculture Organization of the United Nations, Rome. www.fao.org/aq/agl/aglw/aquastat/dbase/index.stm.

Government of India, 2005. Dynamic Ground Water Resources of India. Central Ground Water Board, Ministry ofWater Resources, Government of India, Faridabad, India.
Grey, D., Sadoff, C., 2005.Water Resources, Growth and Development (A working paper for discussion prepared by The World Bank for the panel of Finance Ministers, the UN Commission for Sustainable Development).
Grey, D., Sadoff, C.W., 2007. Sink or swim: water security for growth and development. Water Policy 9 (6), 541–570.
Groundwater Survey and Development Agency/Institute for Resource Analysis and Policy, 2013. Multiple-use Water Services to Reduce Poverty and Vulnerability to Climate Variability and Change Final Report submitted to UNICEF. Institute for Resource Analysis and Policy, Hyderabad.
Haris, A.A., 2013. Integrated farming system for livelihood security of women in water logged areas. In: Meena, M.S., Singh, K.M., Bhatt, B.P., Kumar, U. (Eds.), Gender Perspectives in Integrated Farming Systems. ICAR Research Complex for Eastern Region, Patna, India. A training manual.
Hira, G.S., Khera, K.L., 2000. Water Resource Management in Punjab under Rice-wheat Production System. Research Bulletin No. 2. Department of Soils, Punjab Agricultural University, Ludhiana, India.
Human Development Report, 2006. Human Development Report-2006. United Nations, New York.
Human Development Report, 2009. Human Development Report-2009. United Nations, New York.
Hussain, I., Hanjra, M.A., 2003. Does irrigation water matter for rural poverty alleviation? Evidence from South and South East Asia. Water Policy 5 (5), 429–442.
Jha, N., 2010. Access of the Poor to Water and Sanitation in India: Salient Concepts, Issues and Cases. Working Paper 62. International Policy Centre for Inclusive Growth, Brazil.
van Koppen, B., Smits, S., Moriarty, P., Penning de Vries, F., Mikhail, M., Boelee, E., 2009. Climbing the Water Ladder: Multiple Use Water Services for Poverty Reduction. IRC International Water and Sanitation Centre and International Water Management Institute, Hague, The Netherlands.
Kumar, M.D., Singh, O.P., 2005. Virtual water in global food and water policy making: is there a need for rethinking? Water Resources Management 19 (6), 759–789.
Kumar, M.D., 2007. *Groundwater management in India: physical, institutional and policy alternatives*. Sage Publications, New Delhi.
Kumar, M.D., Ghosh, S., Patel, A.R., Singh, O.P., Ravindranath, R., 2006. Rainwater harvesting in India: some critical issues for basin planning and research. Land Use and Water Resource Research 6 (1), 1–17.
Kumar, M.D., Narayanamoorthy, A., Singh, O.P., Sivamohan, M.V.K., Sharma, M., Bassi, N., 2010. Gujarat's Agricultural Growth Story: Exploding Some Myths. Occasional Paper no. 2. Institute for Resource Analysis and Policy, Hyderabad.
Kumar, M.D., Patel, A., Ravindranath, R., Singh, O.P., 2008. Chasing a mirage: Water harvesting and artificial recharge in naturally water-scarce regions. Economic & Political Weekly 43 (35), 61–71.
Kumar, M.D., 2003. Food Security and Sustainable Agriculture in India: The Water Management Challenge. Working Paper 60. International Water Management Institute, Colombo, Sri Lanka.
Kumar, M.D., 2009. Water Management in India: What Works, what Doesn't? Gyan Books, New Delhi.
Laurence, M., Meigh, J., Sullivan, C., 2002. Water Poverty of Nations: an International Comparison. Kellee Economics Research Papers, Kellee University, Wallingford, UK.
Mc Cully, P., Wong, S., 2004. Powering a sustainable future: the role of large hydropower in sustainable development. In: Paper Prepared for the UN Symposium on Hydropower and Sustainable Development, Beijing, China.

Mc Cully, P., 1996. Climate Change Dooms Dams Silenced Rivers: The Ecology and Politics of Large Dams. Zed Books, London.
Molden, D., Amarasinghe, U., Hussain, I., 2001. Water for Rural Development. Working Paper 32. International Water Management Institute, Colombo, Sri Lanka.
Pearce, D.W., Warford, J.J., 1993. World without End: Economics, Environment, and Sustainable Development. Oxford University Press, London and New York.
Pérez, A.M., Smits, S., Benavides, A., Vargas, S., 2004. Multiple use of water, livelihoods and poverty in Columbia: a case study from the Ambichinte micro-catchment. In: Moriarty, P., Butterworth, J., Van Koppen, B. (Eds.), Beyond Domestic: Case Studies on Poverty and Productive Uses of Water at the Household Level. IRC International Water and Sanitation Centre, Delft, The Netherlands.
Perry, C.J., 2001. World commission on dams: implications for food and irrigation. Irrigation and Drainage 50 (2), 101–107.
Psacharopolous, G., Morley, S., Fiszbein, A., Lee, H., Wood, W.C., 1992. Poverty and Income Distribution in Latin America: The Story of the 1980s. Technical Paper No. 351. World Bank, Washington, DC.
Ramirez, Alegandro, Ranis, G., 1997. Economic Growth and Human Development. Centre Discussion Paper No. 787. Economic Growth Centre, Yale University, New York, US.
Ranis, G., 2004. Human Development and Economic Growth. Discussion Paper No. 887. Economic Growth Centre, Yale University, New York, US.
Renault, D., Wahaj, R., Smits, S., 2013. Multiple Uses of Water Services in Large Irrigation Systems. Auditing and Planning Modernization: The MASSMUS Approach. FAO Irrigation and Drainage Paper 67. Food and Agriculture Organization of the United Nations, Rome.
Rockström, J., Barron, J., Fox, P., 2002. Rainwater management for improving productivity among small holder farmers in drought prone environments. Physics and Chemistry of the Earth-Parts A/B/C 27 (11), 949–959.
Rosenzweig, M.R., 1995. Why are there returns in schooling? American Economic Review 85 (2), 153–158.
Rusere, S., 2005. An Assessment of the Multiple Uses of Small Dams, Water Usage and Productivity in the Limpopo Basin. An Undergraduate Research Project Submitted in Partial Fulfilment of the Requirements of the Degree of Bachelor of Science Honours in Agricultural Engineering, Department of Soil Science and Agricultural Engineering. University of Zimbabwe.
Schultz, T.W., 1975. The value of the ability to deal with disequilibria. Journal of Economic Literature 13 (3), 827–846.
Shah, Z., Kumar, M.D., 2008. In the midst of the large dam controversy: objectives, criteria for assessing large water storages in developing world. Water Resources Management 22 (12), 1799–1824.
Shah, T., van Koppen, B., 2006. Is India ripe for integrated water resources management? Fitting water policy to national development context. Economic and Political Weekly 41 (31), 3413–3421.
Sikka, Alok, 2009. Water Productivity of Different Agricultural Systems. In: Kumar, M.D., Amarasinghe, U. (Eds.), *Water Productivity Improvements in Indian Agriculture: Potentials, Constraints and Prospects*. Strategic Analyses of the National River Linking Project (NRLP) of India- Series 4, International Water Management Institute, Colombo, Sri Lanka.
Smakhtin, V., Revenda, C., Doll, P., 2004. Taking into Account Environmental Water Requirements in Global-Scale Water Resources Assessments Comprehensive Assessment Report 2. International Water Management Institute, Colombo, Sri Lanka.
Sullivan, C., 2002. Calculating water poverty index. World Development 30 (7), 1195–1211.
Tundisi, J.G., Matsumura-Tundisi, T., Tundisi, J.E.M., Abe, D.S., Vanucci, D., Ducrot, R., 2005. Reservoir: Functioning, Multiple Use and Management. Institute International of Ecology, Latin America.

Vora, B.B., 1994. Major and Medium Dams: Myth and the Reality. Hindu Survey of the Environment. The Hindu, Chennai, India.
Wiesmann, David, 2006. A Global Hunger Index: Measurement Concept, Ranking of Countries, and Trends. FCND Discussion Paper 212, International Food Policy Research Institute, Washington DC. December 2006.
Wood, A., 1994. North-South Trade, Employment and Inequality: Changing Fortunes in a Skill-driven World. IDS Development Studies Series. Oxford University Press, Oxford.
World Bank, 2004a. Towards Water Secure Kenya: Water Resources Sector Memorandum. World Bank, Washington, DC.
World Bank, 2004b. World Development Report 2005: A Better Investment Climate for Everyone. World Bank, Washington, DC.
World Bank, 2005. Pakistan Water Economy: Running Dry Report 34081-PK. South Asia Region, Agriculture and Rural Development Unit, World Bank, Washington, DC.
World Bank, 2006a. Managing Water Resources to Maximize Sustainable Growth: A Country Water Resources Assistance Strategy for Ethiopia. World Bank, Washington, DC.
World Bank, 2006b. Hazards of Nature, Risks to Development. An IEG Evaluation of World Bank Assistance for Natural Disasters. Independent Evaluation Group, World Bank, Washington, DC.
World Water Council, 2000. 2nd World Water Forum. (The Hague, The Netherlands).
Yang, H., 2002. Water, Environment and Food Security: A Case Study of the Haihe River Basin in China. http://www.rioc.org/wwf/Water_in_China_Haihe.pdf.

CHAPTER 3

Multiple Water Needs of Rural Households: Studies From Three Agro-Ecologies in Maharashtra

Y. Kabir
UNICEF Field Office, Mumbai, Maharashtra, India

V. Niranjan
Engineering & Research International LLC, Abu Dhabi, United Arab Emirates

N. Bassi
Institute for Resource Analysis and Policy, Delhi, India

M.D. Kumar
Institute for Resource Analysis and Policy, Hyderabad, India

3.1 INTRODUCTION

The rural water supply schemes in India are generally planned to meet the domestic water supply needs of the population. However, rural populations also have many productive water needs (van Koppen et al., 2006; Kumar et al., 2013; Niranjan et al., 2014; Kabir et al., 2015). For instance, households require water for meeting livestock needs, particularly livestock drinking.[1] Likewise, rural households which do not have their own farmland and irrigation sources prefer water for growing vegetables to meet their domestic needs as it is important for the nutritional security of their families. Further, households which are not dependent on agriculture and allied activities for their livelihood may need water for meeting one or more of the productive water needs, such as pottery, fishery, pickle-making, and duck-keeping. This will be important for households that are economically poor (van Koppen et al., 2009; GSDA/IRAP/UNICEF, 2013).

The type of productive water need of a household depends on the cultural background, the agro-climatic setting, and occupational profile of the

[1] However, this is applicable to agricultural families which have no source of irrigation. Families which have sources of irrigation would be able to shift their families and cattle to the farms.

Rural Water Systems for Multiple Uses and Livelihood Security
ISBN 978-0-12-804132-1
http://dx.doi.org/10.1016/B978-0-12-804132-1.00003-2

household (GSDA/IRAP/UNICEF, 2013). As regards the influence of culture, tribal communities in India generally keep small ruminants such as goats; they also raise chickens and undertake backyard cultivation of vegetables. The tribal communities in the northeastern region rear pigs in their homestead. The influence of the agro-climate, in high rainfall, subhumid regions of India, especially in the western Ghat region, prompts farmers to raise fruit trees, especially mangoes and guava, as these trees can grow without irrigation.

Livestock keeping is emerging as a major source of livelihood, even in many low- and medium-rainfall, arid and semiarid regions of India, especially in areas where cereal crops are intensively cultivated, which gives the availability of dry and green fodder (Singh et al., 2004; Kumar, 2007). Farmer-households, which own land with irrigation sources, generally grow vegetables in their fields, and therefore would not need water for productive needs in their dwelling premises. Cattle rearers need water for livestock drinking, unless they have irrigation sources. Families which have regular sources of income from employment are able to manage without water for productive needs, as their economic conditions allow them to buy vegetables and milk from the market (GSDA/IRAP/UNICEF, 2013).

In rural areas, conflicts occur between using water for meeting economic goals and meeting social goals. When water becomes scarce, poor communities often compromise on their personal hygiene requirements in an effort to meet the water requirements for productive needs. Failure on the part of the water supply agencies to maintain supply levels that cover productive as well as domestic needs can result in households not being able to realize the full potential of water as a social good. This can happen for two reasons: (1) available water is reallocated (van Koppen et al., 2009)[2]; and (2) they end up spending a substantial amount of time and effort finding water sources to meet the productive needs, which reduces their ability to fetch water from public systems and use for personal hygiene such as washing, bathing, and sanitation. This generally impacts on their productivity in the long run as the communities become susceptible to water-related diseases.

Another possible outcome is that although the agency has not designed the infrastructure for multiple uses (like cattle drinking,

[2] Systems planned for drinking water and other domestic uses are used for cattle watering, irrigation, and a range of other small-scale productive uses (Lovell, 2000; Moriarty et al., 2004).

homestead irrigation), the system by default becomes a multiple-use system. While some of the unplanned uses may be absorbed by the system, other uses can damage it (van Koppen et al., 2009). This is compounded by the often intermittent and unreliable nature of water supplies. Water supply systems that do not consider the needs of rural communities for sustainable livelihoods, fail to play an important role in their day-to-day life. As they are not able to perform economic activities with the water supplied through public systems, communities show a low level of willingness to pay for their water supply services. This affects the sustainability of the systems as official agencies are not able to recover the costs of their operation and maintenance. Thus a vicious circle is perpetuated.

However, there is a growing appreciation of the fact that whenever such unplanned uses take place from "single-use systems" without causing much damage to the physical infrastructure, it brings about improvements in all four dimensions of livelihood related to water. These dimensions are: freedom from drudgery; health; food production; and income (van Koppen et al., 2009). This leads us to the point that a marginal improvement in drinking water supply infrastructure and a marginal increase in the volume of water supplied could remarkably enhance the value of water supplied in social as well as economic terms. However, as many scholars note, planning of water supply systems for multiple uses is restricted by lack of comprehensive data on the incremental costs and returns (Meinzen-Dick, 1997; van Koppen et al., 2006; van Koppen et al., 2009).[3]

It is generally believed that additional water resources to meet multiple needs would be difficult to find in naturally water-scarce regions. Contrary to this, in most instances, ensuring sustainable allocation of water for the household needs from the available resources, rather than overall availability of water in a locality is a real issue. This is evident from the fact that, at the regional level, the water requirement for household needs would be a small fraction of the total water required, and even in the most water-scarce regions, a huge amount of water is consumed in irrigated production of crops. In view of the above discussions, it is important to identify the different types of water needs of rural households in order to assess the type of rural water systems which should be designed and implemented in different

[3] Renwick et al. (2007) had carried out a broad analysis of the incremental costs and benefits from various levels of MUS, ie, basic-level MUS; intermediate-level MUS; and high-level MUS, covering both new systems and upgrading of existing systems.

agro-climatic and socioeconomic settings that can sustain the water supply for domestic as well as productive needs. The chapter discusses one such study carried out in the Indian state of Maharashtra.

3.2 SELECTION OF VILLAGES FROM THREE AGRO-ECOLOGIES IN MAHARASHTRA

To assess the multiple water needs of rural households, a preliminary survey was undertaken in six villages from three districts of Maharashtra. The first district, Satara, in western Maharashtra, is part of the Western Ghats. The second district, ie, Latur, falls in the Central Plateau area. The third district, ie, Chandrapur, falls in the tribal area of Vidarbha, at low altitude.

From the six villages that were surveyed, three villages (one from each district) were shortlisted using the following physical and socioeconomic criteria: (1) larger proportion of households in the village have good physical access to water for drinking and cooking and other household uses from public systems, either through individual tap connections or through nearby stand posts; (2) the villages are predominantly agrarian, but a significant section of the farm households are not able to meet their farming needs from the available water sources (ponds, tanks, and wells), and therefore demand water for multiple purposes from the public systems, including water for livestock and for kitchen gardens to improve their livelihoods; and (3) the current public water supplies across seasons are less than adequate to meet these needs largely due to competition from agriculture. Nevertheless, conditions are favorable for augmenting the available supplies and improving the physical access, through technological and institutional measures.

Based on these considerations, *Varoshi* in Jawali block of Satara district in Western Maharashtra; *Kerkatta* in Latur block of Latur district in Marathwada; and *Chikhali* in Jiwati block of Chandrapur district in Vidarbha region were selected for the study. These represent three different agro-ecological zones in the state. *Varoshi* is in the high-rainfall zone of western Maharashtra located in the foothills of Western Ghats and is characterized by plenty of local streams flowing down from high altitudes fed by base flows from hilly aquifers. *Kerkatta* is from the low- to medium-rainfall zone of Central Plateau, which is drought-prone and experiences a high degree of aridity with rural water supplies heavily dependent on the limited groundwater resources in the Deccan trap formations. *Chikhali* is from the foot of the hilly forested and assured rainfall zone of Vidarbha but with extremely limited groundwater. All the selected villages face problems of inadequate availability of water for meeting multiple needs, such as animal drinking, vegetable cultivation throughout the year, and water for basic needs during summer months. Quality and reliability of water are not issues in any of the villages. A total of 100 households were surveyed from each selected village.

3.3 SOCIOECONOMIC DETAILS OF RURAL HOUSEHOLDS IN THE SELECTED REGIONS

3.3.1 Average Family Size

In *Varoshi*, the average family size of the households belonging to other backward castes (6.6) was higher than that of the general group (6) and schedule caste (5), whereas in *Kerkatta*, the average family size is highest for the general caste (9), followed by other backward caste communities (8) and lowest for schedule caste families (7). In *Chikhali*, it was highest for schedule tribes (7), followed by nomadic tribes (5), other backward castes (4.5), and schedule caste (4). Overall, family size in *Kerkatta* appears to be larger than those in *Varoshi* and *Chikhali*.

3.3.2 Education Status and Occupational Profile

In *Varoshi*, 66% of females and 90.7% of males are in the below poverty line (BPL) category and 72.4% of females and 90.3% of males belonging to the non-BPL category are literate. In *Kerkatta*, 66% of females and 84.3% of males in BPL category and 83% of females and 94% of males in the non-BPL category are literate. In *Chikhali*, 79% of females and 84% of males in the BPL category and 65% of females and 86% of males in the non-BPL category are literate. These results show that while generally the literacy rate is high, the percentage of people who are literate and with higher qualifications is higher in non-BPL families, and the difference is very significant when they are looked at from a gender perspective (except for female literacy in *Chikhali*). Also, the situation with regard to literacy is much better in *Kerkatta* when compared to *Varoshi* and *Chikhali*.

Concerning occupation, 14–21% of males and 3–44% of females belonging to BPL households are engaged in agriculture as their main occupation. However, in the case of members belonging to non-BPL households, 19–35% of males and 11–58% of females belonging to BPL households are engaged in agriculture. Hence, in all the villages, the proportion of persons engaged in agriculture is higher for non-BPL families.

As regards labor (farm and non-farm labor), a significantly large percentage of the BPL families (ranging from 20 to 32%) earn their income from labor, whereas only 8–10% of the non-BPL families are dependent on wage labor. Hence in all the villages, the proportion of people engaged in wage labor is less for non-BPL families.

3.3.3 Average Land and Livestock Holding

In *Varoshi*, the average landholding size owned by BPL households is 0.46 acres[4] (out of which 89% is cultivated but none of this is irrigated) and for non-BPL households it is 2.21 acres (out of which 50% is cultivated and irrigated). In *Kerkatta*, the average landholding size owned by BPL households is 3.25 acres (out of which 84% is cultivated and 56% is irrigated) and for non-BPL households it is 9.14 acres (out of which 87% is cultivated and 53% is irrigated). In *Chikhali*, the average landholding size owned by BPL households is 5.30 acres (out of which 95% is cultivated but none of this is irrigated) and for non-BPL households it is 8.83 acres (all of this is cultivated but only 6% is irrigated). The data indicate that in *Varoshi*, the average landholding size is much smaller as compared to *Kerkatta* and *Chikhali*. Also, irrigation facilities are least developed in the case of *Chikhali*. Nevertheless, in all the villages, non-BPL households were distinguished by a greater amount of landholding, its cultivation, and irrigation in absolute terms.

Table 3.1 presents average livestock holding of the sample households in the selected villages. The families keep buffaloes, local species of cows, crossbred

Table 3.1 Average livestock holding per household in selected villages

Type	Varoshi	Kerkatta	Chikhali
(A) Indigenous cows			
Milch	1.14	1.00	3.50
Dry	1.14	1.30	1.89
Calf	1.25	1.22	2.50
(B) Crossbred cows			
Milch	1.00	2.00	0.00
Dry	0.00	1.00	0.00
Calf	1.00	2.00	0.00
(C) Buffaloes			
Milch	1.11	1.33	0.00
Dry	1.33	1.73	0.00
Calf	1.42	1.60	0.00
(D) Sheep	0.00	1.00	1.00
(E) Goats	0.00	1.00	2.50
(F) Bullocks	0.00	2.14	2.15

Primary survey of sample households.

[4] One hectare equals about 2.4 acres.

cows, sheeps, goats, and bullocks. In *Kerkatta*, though only a few households have livestock, both the livestock type and average number of livestock per household is highest. Further, families in *Chikhali* mostly own indigenous cows, whereas in *Varoshi* and *Kerkatta* they own both indigenous and crossbred types.

In *Varoshi*, the quantity of green fodder fed to dairy animals (cows, buffaloes, and goats) and draught animals is highest during the monsoon, where plenty of natural grass is available in the area owing to the humid climate, high rainfall, and presence of common land, whereas the quantity of dry fodder fed to animals is highest during the summer months. In *Chikhali*, the quantity of green fodder fed to dairy animals (cows and goats) and draught animals is higher during both the monsoon and summer. In *Kerkatta*, the availability of both green and dry fodder is constrained due to low rainfall and frequent drought events.

3.3.4 Kitchen Gardens in the Selected Villages

In *Varoshi* and *Chikhali*, the climate is favorable for growing vegetables and fruits in kitchen gardens. In *Varoshi*, families grow a wide variety of fruits and vegetables in the area, including mango, jack fruit, papaya, banana, lemon, coconut, arecanut, sapota, french bean, brinjal, chilli, pumpkin, and cucumber. Every family has at least two to three fruit trees in its backyard. In *Chikhali*, mango, papaya, banana, lemon, guava, custard apple, pomegranate, almond, bitter gourd, brinjal, chilly, pumpkin, ladies finger, and cucumber are the major fruits and vegetable crops.

However, in *Kerkatta*, which experiences aridity, very few households maintain kitchen gardens. Even among these, they take it up in very small areas, with a few plants and trees. Unlike what was found in the case of *Varoshi* and *Chikhali*, a smaller number of varieties of fruits and vegetables are found.

3.3.5 Annual Household Income

Table 3.2 shows the range of total annual income from various occupations and income from farming, which includes dairying. In all the villages, a

Table 3.2 Total annual income and farm income for households in selected villages

Village	Total Annual Income Range (INR)	Farm Income Range (INR)
Varoshi	100–254,900	100–103,790
Kerkatta	1080–874,300	2000–706,300
Chikhali	1080–283,306	1080–283,306

Authors' own analysis using primary data.

significant number of households were dependent on agriculture as their main source of livelihood. In *Varoshi* 62 out of 100 families, in *Kerkatta* 49 out of 100 families, and in *Chikhali* 64 out of 100 families were found to have income from farming (including dairying).

3.4 MULTIPLE WATER NEEDS OF RURAL HOUSEHOLDS IN THE SELECTED REGIONS

3.4.1 Characteristics of the Village Water Supply Schemes

The village water supply schemes in the three villages show distinct features. Details on public water supply schemes and sources of water supply (both public and private) in the three selected villages are presented in Table 3.3.

In *Varoshi*, the water supply scheme is based on natural springs. Water from the springs is collected in intermediate surface storage systems (GSR, ie, ground surface reservoirs) through long-distance steel pipes, and then

Table 3.3 Water supply sources in the selected villages

	No of such sources		
Water sources	**Varoshi**	**Kerkatta**	**Chikhali**
(A) Water supply scheme			
Type	Surface water	Groundwater	Groundwater
Water supply Well/ spring	0/1	2/0	2/0
Elevated storage reservoir capacity (L)	40,000	50,000	75,000
GSR capacity (no. and L)	2/7000; 4/3000	2/3000	Nil
Private connections	0 (total 173 households)	108 (total 253 households)	176 (total 337 households)
Stand post	12	1	0
(B) Water sources			
Public wells	1	4	4
Private wells	5	52	15
Working hand pumps	4	1	3
Public bore wells	2	1	0
Private bore wells	7	6	30
Farm ponds	0	2	0
Irrigation ponds	0	8	0
Cement check dam	0	5	9
Earthen check dam	0	0	5

Field survey.

distributed to various smaller tanks located at different locations in different hamlets of the village. There are two GSRs of 7000 L capacity and four GSRs of 3000 L capacity. The small surface storage tanks are provided with taps. Water is also supplied through public stand posts in the streets. The households collect water from these taps. No pressurizing device is used for supplying water to the storage tanks as water is available at a good hydrostatic head by virtue of the height from which the spring originates. In addition to the spring, the village also has two bore wells, connected with pumps, which become the main source of water during the months of summer. There are also four functional public hand pumps in the village.

In the case of *Kerkatta*, the only water supply source is groundwater. Wells are constructed by the village *panchayat*[5] to supply water to the households. The water from the wells is pumped into overhead tanks using motors. Water from the overhead tanks is supplied through pipelines to the individual tap connections provided. There are a total of 108 private tap connections.

In *Chikhali*, like in the case of *Kerkatta*, groundwater is the source of the water supply. Two open dug wells in the crystalline hard rock formations is the main water supply source. Water from the open well is pumped into overhead tanks. From these tanks, water is supplied to individual household tap connections through pipelines. There are 176 individual tap connections in the village.

3.4.2 Households Dependence on Different Sources of Water

In *Varoshi*, the most striking fact is that the number of the households depending on public bore wells (refer to Table 3.4) is much larger than that depending on the common stand post (maximum of 31 families). Another interesting phenomenon is that a large number of households depend on farm wells (of others) during summer months to meet their drinking and cooking needs, other domestic needs, and livestock drinking and homestead gardens (refer to Table 3.4). As a matter of fact, households have to depend on multiple sources of water during these months (at least seven water sources in summer against a maximum of five in other months).

In *Kerkatta*, the largest number of households depends on individual tap connections for drinking water supply and other domestic needs (refer to

[5] Village *panchayat* is a local self-government institution at the village level. Members of the village *panchayat* are elected by the village assembly which includes all the adult citizens of the village.

Table 3.4 Water sources for multiple needs across selected villages

Various water needs	Number of households (only majority) depending on various sources								
	Varoshi			Kerkatta			Chikhali		
	Monsoon	Winter	Summer	Monsoon	Winter	Summer	Monsoon	Winter	Summer
Drinking & cooking	64 (Public bore well)	63 (Public bore well)	85 (Farm well)	58 (Individual tap connection)	58 (Individual tap connection)	55 (Farm well)	63 (Individual tap connection)	63 (Individual tap connection)	63 (Individual tap connection)
Other domestic uses including washing, bathing, and sanitation	65 (Public bore well)	30 (Common stand post)	85 (Farm well)	60 (Individual tap connection)	59 (Individual tap connection)	51 (Farm well)	68 (Individual tap connection)	68 (Individual tap connection)	68 (Individual tap connection)
Homestead gardens	10 (Other sources)	10 (Other sources)	10 (Farm wells and other sources)	11 (Individual tap connection)	12 (Individual tap connection)	9 (Farm well)	7 (Other sources)	12 (Other sources)	11 (Other sources)
Livestock	27 (Public bore well)	28 (Public bore well)	39 (Farm well)	20 (Individual tap connection)	21 (Individual tap connection)	23 (Farm well)	34 (Other sources)	33 (Other sources)	44 (Individual tap connection)
Small-scale enterprise	1 (Common stand post)	1 (Common stand post)	1 (Common stand post & farm well)	1 (Individual tap connection)	1 (Individual tap connection)	1 (Open well)	-	-	-

Authors' analysis based on field survey data.

Table 3.4), followed by hand pumps (varying from 25 in summer to 19 households in other months). Fewer families depend on common stand posts (maximum of 13 households which declines to only seven by the summer months). Furthermore, the number of households depending on private tap connections and stand posts declines significantly during summer months, as the supply dwindles and many more families start depending on farm wells (refer to Table 3.4). For meeting livestock drinking and homestead garden requirements, a larger percentage of the households (which undertake these activities) depend on private tap connections. Here again, their number declines as summer advances, and a larger number of families depend on farm wells (refer to Table 3.4).

In *Chikhali*, like in *Kerkatta*, the number of households depending on individual tap connections is much larger than those dependent on any other available water source (Table 3.4). Also, a large number of households depend on public open wells to meet their drinking and cooking needs (27–30 families), other domestic needs (22–26 families), and livestock drinking requirements (6–13 families). As is the case in *Varoshi*, households depend on multiple sources of water during different seasons (up to seven during the summer months).

To sum up, in *Varoshi* and *Kerkatta*, the water supply from public sources (individual tap connections, common stand posts, and public bore wells) becomes highly unreliable during summer, which is indicated by the greater proportion of families depending on private farm wells. Furthermore, in *Chikhali*, households depend on multiple sources of water during different seasons, which is a clear indication of the scarcity of water from different sources.

3.4.3 Accessibility of the Sources

Table 3.5 presents household accessibility to water sources (time traveled, time spent, and storage facilities) across economic segments during different seasons.

In *Varoshi*, there is a distinct difference between below poverty line (BPL) households and non-BPL households when it comes to tapping difference sources of water for meeting domestic and productive needs. In the case of non-BPL households, there are seven different sources of water to meet domestic and livestock needs, whereas in the case of BPL households, there are three sources of water, ie, common stand posts, public bore wells, and farm wells owned by others. The proportion of families which have access to more than once source is also higher for non-BPL households.

Table 3.5 Accessibility of the water sources in the selected villages

	Varoshi		Kerkatta		Chikhali	
Particulars	BPL house holds	Non-BPL house holds	BPL house holds	Non-BPL house holds	BPL house holds	Non-BPL house holds
(A) Average distance traveled (m)						
Monsoon	254	271	345	433	1356	235
Winter	254	271	323	433	1356	235
Summer	566	1194	1584	2097	1383	340
(B) Average time spent (h)						
Monsoon	0.29	0.32	0.76	0.88	1.13	0.52
Winter	0.29	0.32	0.56	0.66	1.13	0.39
Summer	0.86	1.12	1.21	1.07	1.21	0.57
(C) Average storage facilities available (L)						
Drinking and cooking use	26.4	38.2	31.6	41.8	25.8	27.0
Bathroom and toilet use	298.0	504.4	142.8	268.8	252.5	344.6
Livestock use	130	324.9	39.2	122.5	126.3	141.4

Authors' analysis based on field survey data.

In *Kerkatta*, contrary to the situation in *Varoshi*, the BPL families appear to have better access to water sources (six) in terms of number of households which have access to more than one source and the physical distance to the sources. However, as is the case in *Varoshi*, the non-BPL families depend on a larger number of sources (eight). In *Chikhali*, as is the case with Varoshi, non-BPL families have six different sources of water, whereas BPL families had only four sources of water. Also, the proportion of families which have access to more than one source is also higher for non-BPL households.

Furthermore, the analysis of the average distances traveled by the households reveals that the non-BPL families often traverse large distances to fetch water in *Varoshi* and *Kerkatta*. This does not mean that they have greater hardship, but only shows that they have alternative sources of water to bank on, when there is an acute shortage of water during the summer months. These sources include their own bore wells in the farm and bore wells of neighbors. However, in case of *Chikhali*, it is the BPL families which often traverse large distances to fetch water.

Also, in *Varoshi* and *Kerkatta*, there is a marginal difference in the time spent on fetching water, with more time in the case of non-BPL households. This runs against the conventional wisdom that the poor spend more time collecting water. However, it is important to note that the poor do not have alternative sources to bank on when the public water supply system fails, whereas the rich have alternative sources of water. However, in *Chikhali*, BPL households spent more time collecting water.

Another interesting phenomenon is that both in the case of BPL families and non-BPL families in *Varoshi* and *Kerkatta*, the average distance traveled for fetching water and the time spent for the same increase remarkably during summer, while decreases during winter months. Nevertheless, in all the selected villages, the non-BPL families have better storage facilities for water, which would help tide them over the crisis emerging out of interruptions in water supply.

The above analysis highlights that the poor do not have alternative sources to bank on when the public water supply system fails, whereas the rich do. Hence, merely looking at the time spent in water collection is not an indicator of the hardship. One also has to look at the average amount of water consumed by the different economic segments, like BPL and non-BPL families, which is discussed in the next subsection.

3.4.4 Average per Capita Water Use for Different Purposes of the Households

The average per capita daily water use in the households for different purposes during normal years and drought years is given in Table 3.6. In *Varoshi* and *Kerkatta*, there was not much difference in the volumetric use of water between normal years and drought years, contrary to what one would normally expect. The reason might be that these are basic survival needs which one would not like to compromise, and would find many ways to manage the water to attain the same. However, in *Chikhali*, there was a significant difference in the volumetric use of water between normal years and drought years for drinking and cooking purposes and other domestic uses.

From the point of view of vulnerability, what is important is the wide variation in water use across families not only for other domestic uses, such as washing, bathing, and sanitation but also for some of the very basic survival needs, such as drinking and cooking in all the villages. This can be attributed to the difference in physical access to the water sources, and the source characteristics.

Table 3.6 Average per capita daily water use for different purposes of the households

Villages	Average water requirement/ availability during	Drinking and cooking (L)			Other domestic uses including washing, bathing, and sanitation use (L)		
		Monsoon	Winter	Summer	Monsoon	Winter	Summer
Varoshi	Requirement	6.19	6.19	6.60	23.60	23.60	24.19
	Availability during normal rainfall year	6.13	6.19	6.61	23.60	23.60	24.04
	Availability during drought year	6.18	6.11	6.35	23.65	23.33	23.61
Kerkatta	Requirement	4.70	4.70	5.61	16.74	17.19	17.63
	Availability during normal rainfall year	5.15	4.71	5.56	17.19	17.24	17.73
	Availability during drought year	4.53	4.53	5.31	15.07	15.51	16.13
Chikhali	Requirement	6.32	6.43	8.37	20.19	20.30	20.89
	Availability during normal rainfall year	6.31	6.33	8.44	19.83	20.19	20.65
	Availability during drought year	7.39	7.23	11.22	14.76	12.54	15.38

Authors' own analysis using primary data.

The average daily household level of water use for productive uses in *Varoshi* and *Kerkatta* shows marginal difference in household level water use between normal and drought years for kitchen gardens and livestock drinking. However, in *Chikhali*, there was a significant difference in household level water use between normal years and drought years for kitchen gardens (62 and 20 L, respectively) and livestock drinking (60 and 35 L, respectively). In *Kerkatta*, a difference exists in volumetric water use across seasons, with larger quantities used in summer months for livestock drinking and watering kitchen gardens.

Nevertheless, there was a significant difference in the average household level of water use between the non-BPL and BPL families, with non-BPL families showing higher water use in all subsectors of household water use, such as drinking and cooking, other domestic uses, livestock use, kitchen gardens, and small-scale enterprises in all the villages (Table 3.7). In the case of livestock drinking, the difference could be because of the larger size of the animal holding of the family or family owning livestock types which consume more water (such as buffalo). This difference in water use perhaps explains the larger amount of time spent and distance covered by the

Table 3.7 Daily household (HH) water use for domestic and productive uses in the selected villages

	Varoshi		Kerkatta		Chikhali	
Water use	**BPL HHs**	**Non-BPL HHs**	**BPL HHs**	**Non-BPL HHs**	**BPL HHs**	**Non-BPL HHs**
(A) Domestic water use						
Drinking and cooking (L/HH/day)	28.86	36.09	29.70	35.30	6.66	5.97
Other uses including washing, bathing, and sanitation (L/HH/day)	112.73	140.13	88.3	146.3	17.86	21.80
(B) Productive water use						
Homestead gardens (L/HH/day)	–	33.33	21.5	77.40	52.50	74.44
Livestock (L/HH/day)	43.00	53.30	36.40	81.40	51.55	64.76
Small-scale enterprises (L/HH/day)	–	3000.00	–	60.00	–	–

Authors' analysis based on field survey data.

non-BPL households in water collection and the larger storage facilities available for water at the household level. Also, more importantly, while the non-BPL families use water for homestead gardens, the BPL families do not have them.

3.5 ASSESSING THE VULNERABILITY OF RURAL HOUSEHOLDS TO PROBLEMS ASSOCIATED WITH LACK OF WATER FOR DOMESTIC AND PRODUCTIVE NEEDS

In order to assess the vulnerability of the rural households to inadequate supply of water to meet drinking water, sanitation, and livelihoods needs, a multiple-use water system vulnerability index (MUWSVI) developed by GSDA/IRAP/UNICEF (2013) was used. The MUWSVI comprises of the following six broad subindices: (1) water supply and use; (2) family occupation and social profile; (3) presence of social institutions and ingenuity; (4) water resource endowment; (5) climate and drought proneness; and (6) financial stability. Each subindex has a different set of parameters which influence household vulnerability. For instance, vulnerability reduces with improved access to primary source of water and with the number of alternate sources of water, whereas it increases with a decrease in the required quantity of water and frequency of water delivery and also with longer distances traveled to fetch water from the water sources (all part of water supply and use subindex). Similarly, vulnerability will be low for families with a regular source of livelihood that are not dependent on water, whereas it will be high for families with school-going children or office-going adults (part of family occupation and social profile subindex). Vulnerability will also be high in regions which experience high seasonal variation in water availability (for more details, refer to GSDA/IRAP/UNICEF, 2013; Niranjan et al., 2014; Kabir et al., 2015).

The composite MUWSVI will have a maximum value of 10, meaning zero vulnerability; lower values of the index indicate vulnerability. Households with an index value lower than 5 were considered to be highly vulnerable.

The vulnerability index was computed for all the sample households of the three pilot villages. For computation, several of the natural, physical, social and socioeconomic and financial variables, influencing the vulnerability are considered, with the values of the respective variables being obtained from primary household survey, analysis of primary data, and in a few cases with secondary data. In the case of *Varoshi*, the value of the index was found to vary from 3.31 to 6.58. Fig. 3.1 shows the values for all the sample households. Out of the 100 households surveyed, 67 households have vulnerability index values lower than five.

The computed values of MUWSVI at the household level for the sample households in *Kerkatta* village of Latur are provided in Fig. 3.2. The values range from a low of 2.21 to a high of 6.32. Out of the 100 households, 81 have values lower than 5, and therefore are considered vulnerable from a multiple water use point of view.

The computed values of MUWSVI at the household level for the sample households in *Chikhali* village of Chandrapur district are provided in Fig. 3.3. The values range from a low of 3.15 to a high of 6.37. Out of the 100 households, 30 have a vulnerability index lower than 5, and hence are treated as highly vulnerable.

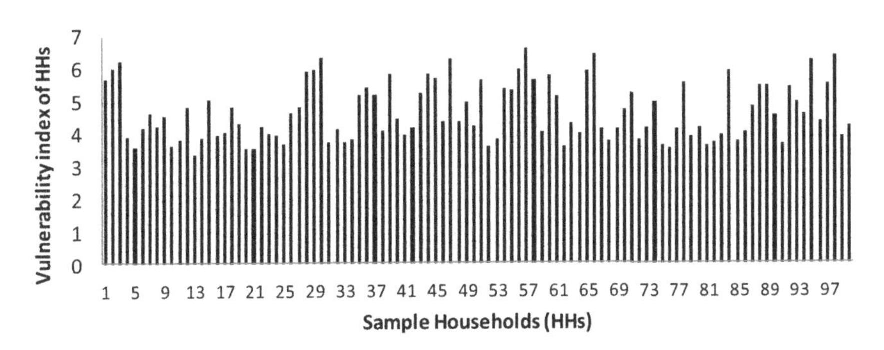

Figure 3.1 Multiple-use vulnerability index for sample households, *Varoshi*, Satara. *Authors' own analysis.*

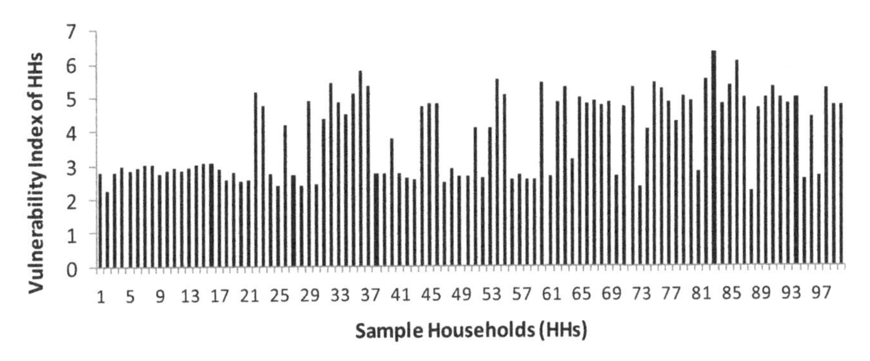

Figure 3.2 Multiple-use vulnerability index for sample households, *Kerkatta*, Latur. *Authors' own analysis.*

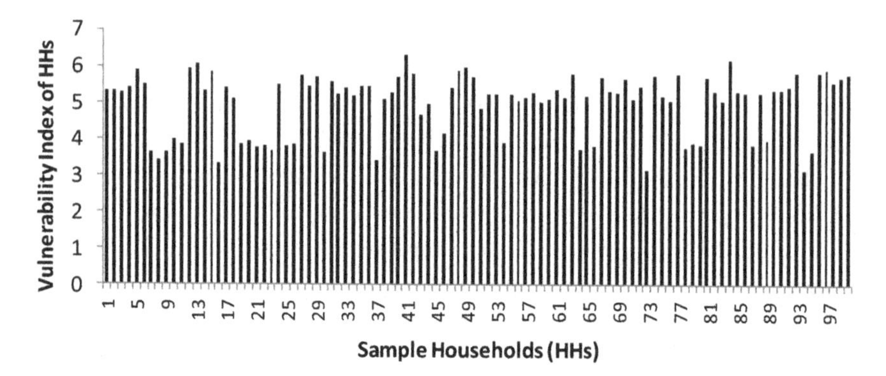

Figure 3.3 Multiple-use vulnerability index for sample households, *Chikhali*, Chandrapur. *Authors' own analysis.*

3.6 CONCLUSION

Analysis of primary data showed that in all three villages, households had domestic and productive water uses, namely, drinking and cooking; bathing, washing, and toilet use; livestock drinking; and, watering of kitchen gardens. Nevertheless, the number of households using water for kitchen gardens and homesteads was much less as compared to other uses. Also, the number of households keeping livestock is lower in *Kerkatta* owing to a shortage of drinking water and green and dry fodder. Furthermore, those who have kitchen gardens, practice it in much smaller areas as compared to *Varoshi*. Again, water shortage is the major reason for this. Only one household each from two villages, namely, *Varoshi* and *Kerkatta*, was found to be using water for small-scale industrial units. In all the villages, water use for domestic and productive needs was found to be influenced by the overall economic condition of the households, with higher values for economically better-off households.

The vulnerability indices for the sample households were computed based on the assessment of the different domestic and productive water needs of the sample households covered during primary survey, and the data obtained on various physical and socioeconomic parameters influencing their MUS vulnerability.

In the case of *Varoshi*, the MUS vulnerability of the HHs varies from 3.31 to 6.58. Sixty-seven families were found to be most vulnerable as per the estimates, with values of the index falling below 5. In spite of being a high rainfall region, *Varoshi* had a significant proportion of vulnerable households which can be attributed to the fact that the main source of water (spring) needs urgent repairs. As a result, villagers were found to be facing water scarcity during summer months and hence higher vulnerability.

In the case of *Kerkatta*, the values range from 2.16 to 6.14. Of these, 81 households were found to be most vulnerable. As *Kerkatta* falls in a low-rainfall and drought-prone area with limited water resource endowment, a higher number of households were found to be vulnerable. In the case of *Chikhali*, the values range from 3.14 to 6.32, and 31 households were found to be most vulnerable. As *Chikhali* is in an assured rainfall region, a comparatively low proportion of households were found to be vulnerable. Overall, *Kerkatta* had the maximum percentage of households vulnerable to problems associated with inadequate water supply for multiple uses.

Based on the assessment of the vulnerability index, interventions to increase access and available quantity of water should be designed (either through water supply augmentation schemes or retrofitting of existing water supply infrastructure) in order to cater to the multiple-use water needs of households. The index will also provide a useful tool for evaluating the impact of new interventions on the vulnerability of households to problems associated with lack of water for multiple needs.

REFERENCES

Groundwater Surveys and Development Agency (GSDA), Institute for Resource Analysis and Policy (IRAP) and UNICEF, 2013. Multiple-Use Water Services to Reduce Poverty and Vulnerability to Climate Variability and Change. Institute for Resource Analysis and Policy, Hyderabad.

Kabir, Y., Vedantam, N., Kumar, M.D., 2015. Women and water: vulnerability from water shortages. In: Cronin, A., Mehta, P.K., Prakash, A. (Eds.), Gender Issues in Water and Sanitation Programmes: Lessons from India. Sage Publications, New Delhi, p. 45.

Kumar, M.D., Panda, R., Vedantam, N., Bassi, N., 2013. Technology choices and institutions for improving economic and livelihood benefits from multiple-use tanks in western Odisha. In: Kumar, M.D., Sivamohan, M.V.K., Bassi, N. (Eds.), Water Management, Food Security and Sustainable Agriculture in Developing Economies. Routledge, London, UK, pp. 138–163.

Kumar, M.D., 2007. Groundwater Management in India: Physical, Institutional and Policy Alternatives. Sage Publications, New Delhi.

Lovell, C., 2000. Productive Water Points in Dryland Areas: Guidelines on Integrated Planning for Rural Water Supply. ITDG Publishing, London, UK.

Meinzen-Dick, R., 1997. Valuing the multiple uses of water. In: Kay, M., Franks, T., Smith, L. (Eds.), Water: Economics, Management and Demand. E&FN Spon, London, UK, pp. 50–58.

Moriarty, P., Butterworth, J., van Koppen, B. (Eds.), 2004. Beyond Domestic: Case Studies on Poverty and Productive Uses of Water at the Household Level. IRC International Water and Sanitation Centre, Delft, The Netherlands (Technical Paper no. 41).

Niranjan, V., Kumar, M.D., Kabir, Y., 2014. Developing a household-level MUWS vulnerability index for rural areas. In: Kumar, M.D., Bassi, N., Narayanamoorthy, A., Sivamohan, M.V.K. (Eds.), The Water, Energy and Food Security Nexus: Lessons from India for Development. Routledge, London, UK, pp. 160–171.

Renwick, M., Joshi, D., Huang, M., Kong, S., Petrova, S., Bennett, G., Bingham, R., Fonseca, C., Moriarty, P., Smits, S., Butterworth, J., Boelee, E., Jayasinghe, G., 2007. Multiple Use Water Services for the Poor: Assessing the State of Knowledge. Winrock International, Arlington, VA.

Singh, O.P., Sharma, A., Singh, R., Shah, T., 2004. Virtual water trade in the dairy economy: analyses of irrigation water productivity in dairy production in Gujarat, India. Economic and Political Weekly 39 (31), 3492–3497.

van Koppen, B., van Smits, S., Moriarty, P., Penning de Vries, F., Mikhail, M., Boelee, E., 2009. Climbing the Water Ladder: Multiple-Use Water Services for Poverty Reduction. Technical Paper series no. 52. IRC International Water and Sanitation Centre and International Water Management Institute, Hague, The Netherlands and Colombo, Sri Lanka.

van Koppen, B., Moriarty, P., Boelee, E., 2006. Multiple-Use Water Services to Advance the Millennium Development Goals Research Report 98. International Water Management Institute, Colombo, Sri Lanka.

CHAPTER 4

Multiple-Use Water Systems for Reducing Household Vulnerability to Water Supply Problems

Y. Kabir
UNICEF Field Office, Mumbai, Maharashtra, India

N. Bassi
Institute for Resource Analysis and Policy, Delhi, India

M.D. Kumar
Institute for Resource Analysis and Policy, Hyderabad, India

V. Niranjan
Engineering & Research International LLC, Abu Dhabi, United Arab Emirates

4.1 INTRODUCTION

In chapter "Multiple Water Needs of Rural Households: Studies From Three Agro-Ecologies in Maharashtra," we identified the multiple water needs of the communities in three different typologies of Maharashtra. Furthermore, we assessed the vulnerability of households to problems associated with lack of an adequate supply of water to meet drinking water, sanitation, and livelihoods needs by deriving a multiple-use water vulnerability index. As per the index values, *Kerkatta* village in Marathwada had the maximum percentage (81%) of households vulnerable to problems associated with inadequate water supply for multiple uses, whereas Chikhali village in Vidarbha had the lowest proportion (31%) of highly vulnerable households. In the case of *Varoshi* village in western Maharashtra, 67% of the households were found to be vulnerable.

This chapter builds on that study and suggests multiple-use water system (MUWS) models for each selected village from the three regions to reduce households' vulnerability to problems of water supply and make them water secure. The model design involved a six-step process which includes: (1) an extensive review of MUWS models from around the world, with particular reference to the physical settings under which they work, and the key design

Rural Water Systems for Multiple Uses and Livelihood Security
ISBN 978-0-12-804132-1
http://dx.doi.org/10.1016/B978-0-12-804132-1.00004-4

features that result in their good performance; (2) study of the physical (hydrology, geology, geo-hydrology, climate, and topography) features of the locations selected for piloting; (3) selection of MUWS models from regions that closely resemble the pilot locations vis-à-vis the physical settings; (4) identification of multiple water needs of the pilot villages based on primary survey (both (2) and (4) have already been discussed in chapter: Multiple Water Needs of Rural Households: Studies From Three Agro-Ecologies in Maharashtra); (5) study of the water supply systems available in the pilot locations, which have the potential to become MUWS; and, (6) implanting the special features of MUWS into the existing systems so as to improve their performance in terms of their ability to meet multiple water needs. Important steps are discussed in the subsequent sections. Furthermore, the institutional setup for management of MUWS has also been proposed.

4.2 EXISTING VILLAGE WATER SUPPLY SYSTEMS IN THREE SELECTED REGIONS

The village water supply schemes in the three villages show distinct features. The water supply scheme in Varoshi is based on natural springs. Water from the springs is collected in intermediate surface storage systems (GSR, ie, ground surface reservoirs) through long-distance steel pipes, and then distributed to various smaller tanks located at different locations in different hamlets of the village. There are two GSRs of 7000 L capacity and five GSRs of 3000 L capacity. The small surface storage tanks are provided with taps. Water is also supplied through public stand posts in the streets. The households collect water from these taps. No pressurizing device is used for supplying water to the storage tanks as water is available at a good hydrostatic head by virtue of the height from which the spring originates. In addition to the spring, the village also has two bore wells, connected with pumps, which become the main source of water during the summer months. There is also one functional public hand pump in the village.

In the case of Kerkatta, the only water supply source is groundwater. Wells are constructed by the village *panchayat*[1] to supply water to the households. The water from the wells is pumped into overhead tanks using motors. Water from the overhead tanks is supplied through pipelines to the individual tap connections provided. All households have private tap

[1] Village *panchayat* is a local self-government institution at the village level. Members of the village *panchayat* are elected by the village assembly which includes all the adult citizens of the village.

connections. There are two public wells, one for drinking and the other for domestic purposes.

In Chikhali, like in the case of Kerkatta, groundwater is the source of the water supply. An open dug well in the crystalline hard rock formations is the main water supply source. Water from the open well is pumped into overhead tanks. From these tanks, water is supplied to individual household tap connections through pipelines. There are 170 individual tap connections in the village. In addition to the open well, there are three functional hand pumps in the village.

4.3 CHARACTERISTICS OF MUWS: FINDINGS FROM A GLOBAL REVIEW

This section summarizes the results of an extensive review undertaken on multiple-use water systems around the world. The results are presented with regard to the physical settings in which the systems work in terms of hydrology, geology, and topography; the description of the MUWS, covering the physical and socioeconomic aspects of water supply and use; and the key features of the MUWS which result in their good performance. A total of 17 different types of MUWS, from countries as far and wide as the flood plains of Bangladesh (Nagabhatla et al., n.d.), Bolivia (van Koppen et al., 2008), Brazil (Tundisi et al., 2005), China (FAO, 2010b), Colombia (Pérez et al., 2004), Ethiopia (van Koppen et al., 2008), India (van Koppen et al., 2008; FAO, 2010a; Khan, 2010), north-eastern Morocco (Boelee and Laamrani, 2004), hills of Nepal (Yoder et al., 2008), South Africa (van Koppen et al., 2008), Sri Lanka (Meinzen-Dick and Bakker, 1999), north-eastern Thailand (de Vries and Ruaysoongnern, 2008), Vietnam (FAO, 2010c), and rural and urban Zimbabwe (Thorpe, 2004; Senzanje et al., 2008) were covered in the review. From within India itself, systems from Karnataka, Maharashtra, and the Indo-Gangetic plains (IGP) were covered.

The results show that the MUWSs display wide heterogeneity in size, basic physical features, and the services they provide. The source of water for MUWS varies from springs (Nepal hills) to roof rainwater catchment system, deep-bore wells, ponds and tanks (northeastern Thailand) to a combination of rainwater-harvesting systems (RWHSs) in medium uplands and secondary reservoirs fed by canal seepage (IGP) to large artificial reservoirs initially built for hydropower (Brazil), offtakes from rivers, and primary, secondary, and tertiary canals of large schemes primarily built for irrigation, drainage, and flood control (Hac Hung Hai, Vietnam, and Shahapur canal in

Karnataka), reservoirs of small communal dams (Zimbabwe), and flood plains (Bangladesh). The size varies from small dams to large reservoirs and canal systems. The broad range of services provided by the systems are irrigation, power generation, flood protection, drainage, fisheries, homestead gardens, cattle drinking, domestic water supply, drinking water, brick-making, recreation, environmental flows, cultural uses, industrial, and municipal water supply.

While some are designed by the agency as multiple-use systems, some have become multiple-use systems by default. In the process, the technical features of some of the systems have undergone changes. In the Nepal hills, water from springs is collected in overhead tanks, and is distributed through two parallel distribution pipes, one for domestic water supply and the other for kitchen gardens. Where a single distribution pipe is provided, two taps are provided to the stand post to avoid conflicts over collection of water. Hence, different water use priorities of the communities are incorporated into the design itself. In the case of the canal systems in the IGP, fish trenches and raised beds were constructed for integrated agriculture and fishery using seepage from canals. Fishes are raised in the ponds, while horticultural crops, vegetables, and pulses are grown on 3-m wide raised bunds. Additional runoff water harvested is used for providing supplementary irrigation to paddies. The seasonally waterlogged area is used for raising paddies and fish using nylon nets. Rainwater-harvesting systems in the plateau regions of uplands irrigate the horticultural crops in their commands.

4.4 DESIGN CONSIDERATIONS FOR MUWS FOR THREE REGIONS OF MAHARASHTRA

The selected three regions, ie, western Maharashtra, Marathawad, and Vidarbha, represent three unique physical settings in terms of topography, agro-climates, geology, and geo-hydrology. The basic principle in the design of the MUWS model is that it increases the effective availability of water for household water supply, in order that the water supply system which is being retrofitted is able to meet various domestic and productive needs of the localities in question year round. While the building of an MUWS can be to raise the performance of an existing water supply system in terms of quality, quantity, and reliability (Renwick, 2008), here an attempt is made to improve the quantity of water supply. The following are the considerations for designing MUS models for the three regions:

1. While determining the additional water supply requirement for meeting multiple water needs of households, an increase in per capita water supply requirement would be considered for only those households which have high vulnerability as per the multiple-use water vulnerability estimates, and accordingly the aggregate demand for the village would be estimated by factoring in this as a percentage of the total sample size into the total population. Here, we treat all households with a vulnerability index below 0.5 as highly vulnerable. The higher the value of the vulnerability index, the lower would be the vulnerability of the household.
2. A multiple-use system can provide varying levels of services—from basic MUWS to an intermediate-level MUWS to a high-level MUWS (van Koppen et al., 2009). The per capita water requirement for meeting multiple needs would be estimated based on a broader consideration of domestic and productive water needs for the majority of the poor village households rather than the specific needs of individual households. Obviously, this would be a function of the agro-climate, local culture, and the traditional occupations of the village households. The agro-climate would not only decide the type of water needs but also determine the demand rates. Traditional vegetable growers would like to have water for kitchen gardens, whereas traditional cattle-rearing communities would like to have water provisions for cattle.
3. Storage augmentation will be only for increasing the water supply potential of the system during the lean season. The lean season being referred to here is the season during which the water supply sources (such as rivers, streams, springs, and aquifers) dry up or start showing low yields. The duration of the lean season will have to be determined on the basis of data from the primary survey of sample households. It is assumed that in the remaining months, only change in operating rules of the water supply system would be able to meet the multiple water needs. This is based on the premise that the domestic and productive water requirement of the households will still be a small fraction of the total water demand (including that in conventional irrigated farming) in any village in India.
4. Improved access to water supply sources is important for the vulnerable households to improve the quantity and reliability of water supplies without much effort. For this, the physical distance of the poor households to water supply points (taps, stand posts) should be minimized. The most ideal is household connections.

5. The water requirement for kitchen gardens is assumed to be 75L per family per day (van Koppen et al., 2009). Water requirements for livestock are taken from Singh et al. (2004). Minor corrections for climate can be introduced in this, providing lower values for cold and humid climates.

4.5 MUWS MODELS FOR DIFFERENT REGIONS OF MAHARASHTRA

4.5.1 Western Maharashtra

The MUWS system being proposed for the first region (western Maharashtra) is one which harnesses the natural springs in the area. The discharge of the springs in the area is highly variable, with maximum discharge during the monsoon season, which slowly reduces toward the winter and summer and eventually drying up toward the month of April. The water needs during summer months are met from bore wells in the area, and scarcity is felt. An underground reservoir is proposed to be built in the village, which is capable of diverting and storing water from the "flowing spring" during monsoon months. This should completely prevent evaporation, which would be high during summer months.

The volumetric storage capacity of this reservoir shall be fixed in a such a way that the stored water and the water pumped from the bore wells during the two summer months is sufficient to meet the domestic and productive water needs of the entire population during the two months of summer. As the spring discharge is very high and the domestic demands are relatively low during the rainy season, the diversion of water from the spring into an additional reservoir (other than the intermediate storage systems in the village built for water supply) does not adversely affect the water supply performance during the months of monsoon.

The reservoir and the bore wells can directly feed water into the intermediate storage tanks existing in the village for distribution among the clusters, and individual tap connections of the households. Here in this case, the productive water needs would include water for livestock drinking, water for irrigating the fruit trees, and water for kitchen gardens. This is based on the existing productive water needs of the communities in the village. They were found to be raising trees in their homesteads, and raising kitchen gardens, as well as dairy production.

In *Varoshi* village, since the poor households do not have the independent tap connections and are dependent on common stand posts, it is important to create a water distribution infrastructure, which includes individual tap connections. This would enable the poor households to raise kitchen gardens, apart from covering their human and livestock needs.

4.5.2 Marathwada Region

In the Marathwada village, the communities are entirely dependent on groundwater abstracted from bore wells. The surface water resources are extremely limited in this region. The traditional ponds and tanks in the region dry up completely by the end of winter. The shallow aquifer, which is of basalt origin, has poor groundwater potential. However, after the rains, the water level rises significantly, and often the aquifer does not have sufficient storage capacity to accommodate the water which infiltrates. Overflowing of wells in years of good monsoon is a common phenomenon. But with the withdrawal of water soon after the monsoon, water levels start declining very rapidly. Besides pumping for agriculture, one of the major sources of discharge of water from the shallow aquifer is the outflow into streams and rivers. This basically means that the domestic water sector faces severe competition from irrigation.

The best way for effective allocation of water from a common pool like an aquifer for domestic uses during the lean season is to create pumped storage of the water from the aquifer. Here again, like in the case of western Maharashtra, water required to meet demand for two months can be pumped out when the aquifer is fully replenished. This allows for further infiltration of rainwater, though this would mean reduced stream flows into local ponds and streams and rivers. The water pumped out of the aquifer through bore wells can be stored in surface/subsurface storage tanks. The tank can be lined using HDPE (high-density polyethylene) sheeting to prevent seepage and percolation of the stored water.

The storage capacity of the tank can be worked out in such a way that it takes care of the entire household water demand for domestic and productive needs. Hence, it will be on the basis of: (1) the size of the population; (2) the per capita water demand for domestic and productive uses; (3) the number of months for which the water shortage is felt; and (4) the amount of water tapped daily from the two public open wells during the months of shortage. However, unlike in the case of *Varoshi*, the productive water needs of the population here would cover only water for cattle drinking. It is to be kept in mind here that dairy production is a major source of livelihood and income for the majority of the village households. This essentially means that the greater the degree of water deficit (between the demand during the lean season and the amount supplied though open wells), the higher would be the storage capacity required for the tank.

The only design variable is the total number of bore wells, which during the monsoon can sustainably yield the water, which is needed during the lean season. Since the water table would be very high during monsoon with no significant abstraction for agriculture, the well yield would generally be high.

4.5.3 Vidarbha Region

Chandrapur district of Vidarbha region is characterized by moderate to high annual rainfall (1050 mm), steep slopes, high-density drainage, and hard rock geology. The region forms the upper catchment of the only water-rich river basin flowing in the South Indian peninsula, ie, Godavari. However, the public water supply systems built in the villages of this region are based on groundwater resources, which are extremely scarce due to the crystalline geology. The wells in this area go dry in summer, and there is an acute shortage of drinking water. The only way to augment water supplies for domestic purposes is to harness the surface runoff generated in the region and store it on the surface. Storing water underground won't make much sense for three important reasons. First, recharging the aquifer would be extremely difficult because of the crystalline geology. Second, once water percolates down to the local aquifer, the agricultural users would be the main beneficiary of the recharge water and it would be difficult to enforce any control over agricultural pumpers, with the result that water will be made available for use in summer months; and third, even if agricultural users do not pump out the recharged water, the water might not remain in the local geological formations due to high groundwater flow gradients and outflow of groundwater into the streams.

The water from small hilly catchments in the forest area can be harnessed using small dams and reservoirs. The storage efficiency would be good because of crystalline strata. This water can be pumped into the overhead reservoirs along with the groundwater during the summer months to supply to the individual household connections. Since there would be excessive stream flow during monsoon months, one of the ways to increase the effective storage of the small dam would be to also divert water from the small reservoir during the times of inflow. This strategy can reduce the amount of groundwater pumped from the local aquifer, thereby reducing the pumping costs.

The actual storage capacity (live storage) of the small dam can be estimated on the basis of the following: (1) the size of population in the village; (2) the per capita water demand for domestic and productive uses; (3) the number of months for which the water shortage is felt; and (4) the amount of water tapped daily from open wells and hand pumps during the months of shortage. It is to be kept in mind that the productive water needs of the population here would include water for livestock (cattle, buffaloes, goats, and sheep) drinking and kitchen gardens. This essentially means that the greater the degree of water deficit (between the demand during the lean season and the amount supplied though bore wells), the higher would be the storage capacity required for the tank.

The number of small dams to be constructed will have to be decided on the basis of the maximum storage that can be created at a site, the total storage requirement, and the number of economically viable sites available in the vicinity of the village settlements. The storage capacity of the reservoirs created by the small dams will have to be decided by the rate of runoff (dependable) from the catchment and the catchment area. Water from the small dams can be pumped into the overhead tanks, which can then be distributed among the households through the individual tap connections, which already exist.

4.6 CHARACTERISTICS OF EFFECTIVE MICRO-LEVEL INSTITUTIONS FOR WATER: FINDINGS FROM A REVIEW

4.6.1 The Review of Micro-Level Water Institutions

A review on a wide variety of institutions from different parts of India, managing water-related services was undertaken (includes work by Bardhan, 1993; Mishra, 1996; Meinzen-Dick and Zwarteveen, 1998; Easter et al., 1999; Saleth and Dinar, 1999; Kumar, 2000, 2009, 2010; Isham and Kahkonen, 2002; Saravanan, 2002; Kurian, 2004; Nisha, 2005; Saleth, 2005; Sangameswaran, 2006; van Koppen et al., 2006; Bhavsar and Bhalge, 2007; Meinzen-Dick, 2007; Mikhail and Yoder, 2008; Bassi et al., 2010). These institutions are broadly classified as "local management institutions," and their management regime is either a village, village service area, or area under the distributory outlet or minor outlet of a canal irrigation system; command area of a diversion weir (as in the case of *phad*[2] in Maharashtra)

[2] *Phad* system is a very old and traditional system of gravity-led irrigation through weirs on stream and rivers.

or the command area of a tube well (Mehsana district of north Gujarat); or service area of a village drinking water supply scheme (as in the Indian states of Kerala, Karnataka, and Maharashtra); or the subsystem of a regional water supply scheme, serving a village (Saurashtra in Gujarat); or a watershed area within a village (Ahmednagar in Maharashtra, Uttarakhand, and Himachal Pradesh); or the command area of a surface irrigation tank (Maharashtra, Andhra Pradesh, and Tamil Nadu).

The nature of institutions also varies significantly—from community-based organizations (as found in the case of watershed institutions and *pani samities*[3] for rural drinking water supply), to organizations of landed farmers (as found in the case of groundwater irrigation organizations, water user associations of canal and tank commands) to organizations of both landed and landless farmers (in the case of *pani panchayats*[4]) to irrigation department in the case of *phad* in Maharashtra. The water user associations (WUAs) in canal commands of many states are legal entities registered under the Societies Act. The groundwater irrigation organizations are informal organizations.

The *pani samities*, promoted by the non-government organizations (NGOs) under the aegis of the government agencies (as found in Kerala, Karnataka, and Gujarat) are legitimate organizations, recognized by the village *panchayats*. The watershed institutions (watershed committees) are also legal entities, recognized by the rural development department for undertaking various watershed management activities. The WUAs under tank commands of South India enjoy great legitimacy in the traditional village *panchayats*, but many of the new ones are promoted by the state minor irrigation departments with the help of NGOs and are legally registered.

Performance of the various microinstitutions has been analyzed vis-à-vis their outcomes and impacts on access to water, food security, and livelihoods. Based on the performance of these microinstitutions, and a review of institutional design principles prescribed by scholars working on institutional aspects of sustainable water management, ideal intuitional characteristics that can be introduced into the local MUWS institution have been identified.

4.6.2 The Characteristics of Ideal Institutions MUWS Management

The ideal institutional characteristics for sustainable resource management and multiple water use services are identified from an extensive review of

[3] *Pani samitis* are the local institutions in Gujarat to plan, implement, manage, own, operate, and maintain village water supply systems.

[4] *Pani panchayat* is a voluntary group of farmers engaged in the collective management of water.

the range of effective micro-level water institutions in India, with the purpose of understanding the key design features that contribute to their robustness; and, institutional design principles for sustainable water management prescribed by scholars internationally (see for instance, Frederiksen, 1998). They are as follows:

1. Intersectoral allocation of water from the river basin should be the responsibility of the river basin organization (RBO), whose functioning is regulated by a state-level water resources regulatory agency (in this case Maharashtra State Water Resources Regulatory Authority).
2. The various agencies engaged in managing water-related services (line agencies) such as irrigation and water supply in rural and urban areas would be allowed to appropriate water from within the basin on the basis of the basin-wide allocation for respective sectors.
3. Apart from enforcing water allocation, the RBO will also undertake a water resource management function, including water quality management.
4. Size of MUWS should be such that it can be easily managed by the village community. The village community chooses the location, design, and size of the MUWS on the basis of their needs and priorities.
5. Framing operational rules for the MUWS, including rules for allocation of water across different segments of the socioeconomic strata must be the responsibility of the local community institutions managing MUWS.
6. Drinking water and water for personal hygiene should be given top priority in water allocation from MUWS.
7. Access equity should be the underlying principle for water allocation across households.
8. The local institution managing the MUWS will decide the levels of service, ie, the per capita supply norms, the quality of water; and the frequency and duration of supply.
9. Water prices to be fixed by the local institution need to reflect the scarcity value of the resource to encourage efficient use. Nevertheless, it should be sufficient to cover the cost of operation and maintenance (O&M), including the management costs. The prices being charged from the users should reflect the volumetric consumption.
10. The local MUWS institution needs to be an autonomous entity, with representation of all primary stakeholders of water, for better coordination among users, and management of MUWS.

11. Individuals with good leadership qualities and integrity need to be the entrusted with the responsibility of governance of the institutions at the local level.

4.7 INSTITUTIONAL SET UP FOR MANAGEMENT OF MUWS

As per the Maharashtra Water Resources Regulatory Authority Act (2005), River Basin Agencies issue bulk water entitlements based on the category of use (irrigation water supply, rural water supply, municipal water supply, or industrial water supply) and subject to the priority assigned to such use under State Water Policy. Further, it was stipulated that the existing Irrigation Development Corporations (IDC) should function as River Basin Agencies. However, currently IDCs are only responsible for survey, planning, design, construction, and management of major, medium and minor irrigation projects, and irrigation water services. Hence, the new arrangement is likely to protect the entrenched interests of irrigation corporations to overallocate water for the sector it is concerned with, which is against the principle of sustainable water resources management in the river basin. Researchers for a long time have argued that for sustainable water resource management, the agency responsible for water allocation should not be same as the agency using the water (Frederiksen, 1998; Kumar, 2006). Separating out the water resource management functions from water service functions would create conditions under which the utility (the water supply agency) would be confronted with the opportunity cost of using the water (Arghyam/IRAP, 2010).

Thus, we propose the RBO be responsible for regulating water resource development, performing water resource management functions, managing intersectoral allocation, as per the regulatory framework provided by the Maharashtra Water Resources Regulatory Authority Act (Fig. 4.1). The concerned RBO can share the basin water allocation plans with all concerned departments including the Water Resources Department (WRD) and Water Supply & Sanitation Department (WSSD) and Groundwater Surveys and Development Agency (GSDA), so that planning of schemes by these utilities adheres to such allocation plans. The GSDA and district council should make sure that there is no violation of water rights by putting effective regulatory mechanisms in place. The RBO shall charge for bulk water allocation to various water utilities on the basis of volume supplied to cover the resource cost. The

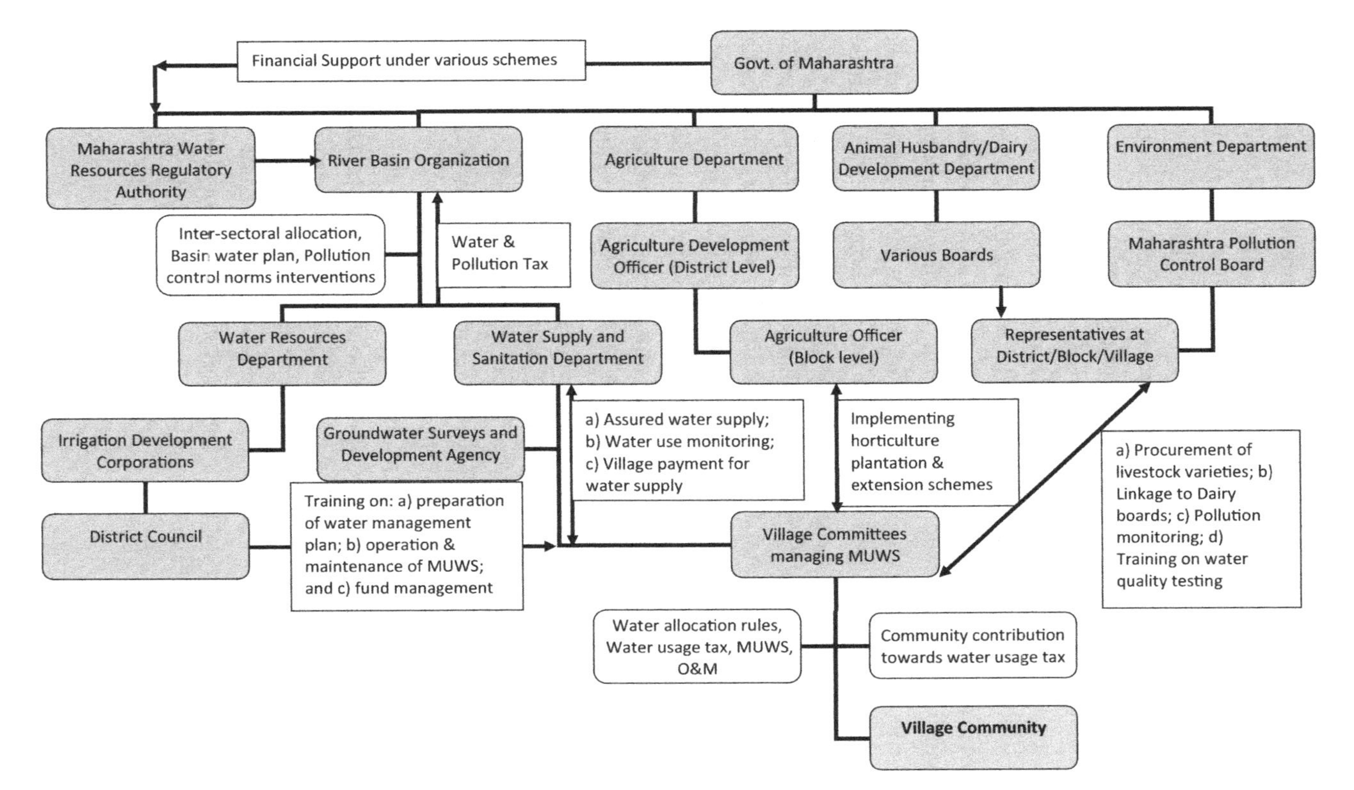

Figure 4.1 Suggested institutional structure for MUWS management in Maharashtra.

presence of formal water markets should encourage the RBOs to manage the resource efficiently and sustainably (Sibly and Tooth, 2007).

Under this institutional paradigm, pollution monitoring can rest with the State Pollution Control Authority, while enforcement of pollution control norms, and water quality management would rest with RBOs. The institutional design principle being followed here is that the agency responsible for monitoring pollution and the agency enforcing pollution control norms, including treatment measures, are not the same.

The GSDA, which is concerned with rural drinking water supply, should guarantee that the village community is able to access the required amount of water, by investing in the necessary infrastructure for diversion/storage and transportation of water, wherever necessary. Since in the case of MUWS, water is also being supplied for uses other than drinking water supply, the cost will have to be shared by concerned agencies, such as the GSDA, animal husbandry department, the fisheries department, the horticulture department, etc. depending on the actual situation in the field vis-à-vis water services.

In return, local MUWS institutions shall pay for the water services, which cover the cost of production and supply of water in addition to the resource cost. At village level, local MUWS institutions should frame operational rules for the MUWS, including rules for allocation of water across different segments and pricing or tax structure for the water services. The water price or tax should reflect the volumetric consumption. Since metering may not be a viable option in the village water supply due to the small volumes of water handled, the local institution can evolve some simple mechanisms for estimating the water drawn by individual households from the system.

Local MUWS institutions will also be responsible for water quality testing at the point of use. For the purpose, required training can be provided by the block/village-level representatives of Maharashtra Pollution Control Board. The GSDA and district council should also arrange the required number of training (related to O&M of MUWS, village water plan, etc.) for the smooth functioning of local MUWS institutions. Considering the proposed multiple use of water (also for livestock and kitchen garden purposes), block/village-level representatives from agriculture, horticulture, and the animal husbandry department should be involved to provide the necessary extension and support services.

4.8 CONCLUSIONS

The design of MUWS involved an extensive review of MUWS models from around the world, which included countries as diverse as India (south and north India), Nepal, Bangladesh (south Asia); Ethiopia, South Africa, Zimbabwe, and Morocco (Africa); Bolivia, Colombia, and Brazil (South America); and Vietnam and Thailand (southeast Asia). The MUWS displayed wide heterogeneity vis-à-vis their technical features, size, and the services they provide. Based on the identification of features which contribute to their effectiveness as MUWS, understanding of the physical and socio-economic characteristics of the regions in question and the type of water supply systems existing there, models of MUWS were developed for all three locations. Subsequently, the institutional arrangements for their management were also worked out. The suggested institutional structure involves a four-tier hierarchy of institutions from the state water resource regulatory authority to the RBOs to the line agencies concerned with water supply services to the local village level institution for managing multiple water use services.

Slightly modified versions of the proposed MUWS models, adjusted mainly to make them financially more viable, have already been executed in the villages of *Varoshi*, *Kerkatta*, and *Chikhali*. A quick estimation of the household vulnerability index post the interventions showed some reduction in the vulnerability of the households to problems associated with lack of water for multiple needs. The highest reduction (28%) was found in the case of *Kerkatta* village, which had the highest number of vulnerable households (81) prior to the intervention, whereas the lowest was in Chikhali where the number of highly vulnerable households reduced by merely 3%, ie, from 30 to 29. In *Varoshi*, the number of highly vulnerable households came down from 67 to 58. Thus, overall the proposed models were found to be effective in improving the water security of the rural households. However, to ascertain the real usefulness of the interventions in improving the household water supply situation and thereby reducing the household vulnerability, impact assessments should be undertaken during summer months when the water sources, which are augmented by the technical interventions suggested under the MUWS models, are dried up.

REFERENCES

Arghyam and Institute for Resource Analysis and Policy (IRAP), 2010. Developing a Tool Kit on Integrated Urban Water Management in India. Institute for Resource Analysis and Policy, Hyderabad.

Bardhan, P., 1993. Analytics of the institutions of informal cooperation in rural development. World Development 21 (4), 633–639.

Bassi, N., Rishi, P., Choudhury, N., 2010. Institutional organizers and collective action: the case of water users' associations in Gujarat, India. Water International 35 (1), 18–33.

Bhavsar, C., Bhalge, P., 2007. Fruits of participatory irrigation management. In: Paper Presented at 4th Asian Regional Conference & 10th International Seminar on Participatory Irrigation Management, Tehran, Iran.
Boelee, E., Laamrani, H., 2004. Multiple use of irrigation water in Northeastern Morocco. In: Moriarty, P., Butterworth, J., van Koppen, B. (Eds.), Beyond Domestic: Case Studies on Poverty and Productive Uses of Water at the Household Level. IRC International Water and Sanitation Centre, Natural Resources Institute and International Water Management Institute, Delft, The Netherlands, pp. 119–135.
de Vries, F.P., Ruaysoongnern, S., 2008. Multiple sources and uses of water in northeast Thailand. In: Proceedings of the CGIAR Challenge Program on Water and Food 2nd International Forum on Water and Food, Addis Ababa, Ethiopia, pp. 83–86.
Easter, K.W., Rosegrant, M.W., Dinar, A., 1999. Formal and informal markets for water: Institutions, performance, and constraints. The World Bank Research Observer 14 (1), 99–116.
Food and Agriculture Organization (FAO), 2010a. Mapping Systems and Service for Multiple Uses in Shahapur Branch Canal Upper Krishna Project KJBNL, Karnataka, India. Land and Water Division, FAO, Rome.
Food and Agriculture Organization (FAO), 2010b. Mapping Systems and Service for Multiple Uses in Fenhe Irrigation District Shanxi Province, China. Land and Water Division, FAO, Rome.
Food and Agriculture Organization (FAO), 2010c. Mapping Systems and Services for Multiple Uses in Bac Hung Hai Irrigation and Drainage Scheme, Vietnam. Land and Water Division, FAO, Rome.
Frederiksen, H.D., 1998. Institutional principles for sound management of water and related environmental resources. In: Biswas, A.K. (Ed.), Water Resources: Environmental Planning, Management, and Development. Tata McGraw-Hill, New Delhi.
Isham, J., Kahkonen, S., 2002. Institutional Determinants of the Impact of Community-based Water Services: Evidence from Sri Lanka and India. Economics Discussion Paper no. 02-20. Department of Economics, Middlebury College, Middlebury, Vermont.
Khan, M.A., 2010. Enhancing water productivity through multiple uses of water in Indo-Gangetic basin. In: Proceedings of the Conference on Innovation and Sustainable Development in Agriculture and Food, Montpellier, France, p. 10.
Kumar, M.D., 2000. Institutions for efficient and equitable use of groundwater: Irrigation management institutions and water markets in Gujarat, Western India. Asia-Pacific Journal of Rural Development 10 (1), 53–65.
Kumar, M.D., 2006. Water Management in River Basins: A Case Study of Sabarmati River Basin in Gujarat (Ph.D. thesis). School of Business Management, Sardar Patel University, Anand, Gujarat.
Kumar, M.D., 2009. Water Management in India: What Works, what Doesn't? Gyan Books, New Delhi.
Kumar, M.D., 2010. Managing Water in River Basins: Hydrology, Economics, and Institutions. Oxford University Press, New Delhi.
Kurian, M., 2004. Institutions for Integrated Water-Resources Management in River Basins: An Analytical Framework. Working Paper 78. International Water Management Institute, Colombo, Sri Lanka.
Meinzen-Dick, R., Bakker, M., 1999. Irrigation systems as multiple-use commons: water use in Kirindi Oya, Sri Lanka. Agriculture and Human Values 16 (3), 281–293.
Meinzen-Dick, R., Zwarteveen, M., 1998. Gendered participation in water management: Issues and illustrations from water users' associations in South Asia. Agriculture and Human Values 15, 337–345.
Meinzen-Dick, R., 2007. Beyond panaceas in water institutions. PNAS 104 (39), 15200–15205.

Mikhail, M., Yoder, R., 2008. Multiple-Use Water Service Implementation in Nepal and India: Experience and Lessons for Scale-Up. International Development Enterprises and Challenge Program on Water and Food, International Water Management Institute, USA and Sri Lanka.
Mishra, B., 1996. A successful case of participatory watershed management at Ralegan Siddhi village in district Ahmadnagar, Maharastra, India. In: Sharma, P.N., Wagley, M.P. (Eds.), Case Studies of People's Participation in Watershed Management in Asia. PWMTA Program, Food and Agriculture Organization, Kathmandu, Nepal.
Maharashtra Water Resources Regulatory Authority Act (MWRRAA), 2005. Law, Environment and Development Journal 1 (1), 80–96.
Nagabhatla, N., Nguyen-Khoa, S., Beveridge, M., Haque, A.B.M., Sheriff, N., Van Brakel, M., Rahman, F., Barman, B., n.d. Multiple-Use of Water in Bangladesh Floodplains: Seasonal Aquaculture and Conjunctive Use of Surface and Groundwater for Improved Rice-fish Production Systems. The World Fish Center, Penang, Malaysia.
Nisha, K.R., 2005. Institutional arrangements in rural water supply in Kerala: constraints and possibilities. In: Paper Presented at 4th IWMI-Tata Annual Partners Meet. Anand, Gujarat.
Pérez, M.A., Smits, S., Benavides, A., Vargas, S., 2004. Multiple use of water, livelihoods and poverty in Colombia: a case study from the Ambichinte micro-catchment. In: Moriarty, P., Butterworth, J., van Koppen, B. (Eds.), Beyond Domestic: Case Studies on Poverty and Productive Uses of Water at the Household Level. IRC International Water and Sanitation Centre, Natural Resources Institute and International Water Management Institute, Delft, The Netherlands, pp. 75–93.
Renwick, M., 2008. Multiple Use Water Services. Winrock International, Nairobi, Kenya.
Saleth, R.M., Dinar, A., 1999. Evaluating Water Institutions and Water Sector Performance. Technical Paper no. 447. World Bank, Washington, DC, USA.
Saleth, R.M., 2005. Water institutions in India: structure, performance, and change. In: Gopalakrishnan, C., Tortajada, C., Biswas, A.K. (Eds.), Water Institutions: Policies, Performance, and Prospects. Springer, New York, pp. 47–80.
Sangameswaran, P., 2006. Equity in watershed development: a case study in western Maharashtra. Economic and Political Weekly 41 (21), 2157–2165.
Saravanan, V.S., 2002. Institutionalizing community-based watershed management in India: elements of institutional sustainability. Water Science and Technology 45 (11), 113–124.
Senzanje, A., Boelee, E., Rusere, S., 2008. Multiple use of water and water productivity of communal small dams in the Limpopo Basin, Zimbabwe. Irrigation and Drainage Systems 22 (3–4), 225–237.
Sibly, H., Tooth, R., 2007. Bringing Competition to Urban Water Supply. School of Economics and Finance, UTAS, Australia.
Singh, O.P., Sharma, A., Singh, R., Shah, T., 2004. Virtual water trade in the dairy economy: analyses of irrigation water productivity in dairy production in Gujarat, India. Economic and Political Weekly 39 (31), 3492–3497.
Thorpe, I., 2004. Pump Aid and the Elephant Pump in Zimbabwe. A Case study for E-conference: Tackling poverty through multiple use water services Available at: www.musgroup.net/content/download/188/1601/file/pumpaid.pdf (accessed 21.09.12.).
Tundisi, J.G., Matsumura-Tundisi, T., Tundisi, J.E.M., Abe, D.S., Vanucci, D., Ducrot, R., 2005. Reservoirs: Functioning, Multiple Use and Management. Agriculture Research for Development (CIRAD), Paris.
van Koppen, B., van Smits, S., Moriarty, P., Penning de Vries, F., Mikhail, M., Boelee, E., 2009. Climbing the Water Ladder: Multiple-Use Water Services for Poverty Reduction. TP series no. 52. IRC International Water and Sanitation Centre and International Water Management Institute, Hague, The Netherlands and Colombo, Sri Lanka.
van Koppen, B., Moriarty, P., Boelee, E., 2006. Multiple-use Water Services to Advance the Millennium Development Goals Research Report 98. International Water Management Institute, Colombo, Sri Lanka.

van Koppen, B., van Smits, S., Moriarty, P., Penning de Vries, F., 2008. Community-level multiple-use water services: MUS to climb the water ladder. In: Fighting Poverty through Sustainable Water Use: Proceedings of the CGIAR Challenge Program on Water and Food 2nd International Forum on Water and Food, vol. 2. The CGIAR Challenge Program on Water and Food, Colombo, Sri Lanka, pp. 217–221.

Yoder, R., Mikhail, M., Sharma, K., Adhikari, D., 2008. Technology adoption and adaptation for multiple use water services in the hills of Nepal. In: Proceedings of the CGIAR Challenge Program on Water and Food 2nd International Forum on Water and Food, Addis Ababa, Ethiopia, pp. 99–102.

CHAPTER 5

Sustainability Versus Local Management: Comparative Performance of Rural Water Supply Schemes

N. Bassi
Institute for Resource Analysis and Policy, Delhi, India

Y. Kabir
UNICEF Field Office, Mumbai, Maharashtra, India

5.1 INTRODUCTION

Ever since independence, India had invested a whooping sum of nearly US$ 11,330 million[1] towards improving water supplies in rural areas through various schemes and programs during the plan periods until 2007. In fact, in the late 1990s, the whole approach to the provision of rural water supply was changed from "supply-driven," which had more emphasis on construction and creation of assets, to "demand-driven," where community participation and decentralization of powers for implementing and operating drinking water supply was made the essential point. Furthermore, the government role was restricted to that of a facilitator.

These efforts aimed at improving the coverage of the schemes in terms of number of villages and habitations. Though significant achievements were made in terms of improving the coverage, they do not commensurate with the scale of investments that have gone into the sector owing to the increasing number of habitations slipping back to the "no-source category." This "slip back" is because of the several threats the rural water supply sector is facing which can be summarized as: poor sustainability of water supplies owing to unsustainable resource base; soaring operation and maintenance costs due to the absence of regular upkeep, and poor cost recovery, both resulting in poor financial working of the system; and inequity in access to

[1] Presently, 1US$ is equal to about 65 Indian National Rupee (INR).

Rural Water Systems for Multiple Uses and Livelihood Security
ISBN 978-0-12-804132-1
http://dx.doi.org/10.1016/B978-0-12-804132-1.00005-6

water across different segments owing to lack of adequate institutional capacities built at the local level for water distribution.

Furthermore, very little information is available on how effective this expenditure has been in providing safe water to rural populations (World Bank, 2008). Also, there is hardly any analysis of the cost of water supply schemes, cost recovery and subsidies, and the impact of technology choice and institutional arrangements on the level of service. Pattanayak et al. (2007) also observed that there were only a few rigorous scientific impact evaluations showing the effectiveness of rural water supply policies in delivering many of the desired outcomes. Their analysis showed that such policies are complex, with multiple objectives; use inputs from multiple sectors; provide a variety of services (water supply, water quality, sanitation, sewerage, and hygiene) using a variety of types of delivery (public interventions, private interventions; public–private partnerships, decentralized delivery, expansion or rehabilitation); and generate effects in multiple sectors (water, environment, health, labor). The fact that the communities are mostly dependent on multiple sources of water supply, including informal sources, often makes the evaluation of policy impacts complicated.

As an outcome, even after making remarkable progress with respect to provision of rural water supply services in India, many habitations and villages continue to face water scarcity during summer months. The problem is more pronounced in areas underlain by hard rocks and where the water supply is from wells. In such areas, due to the constraints imposed by the geo-hydrological settings (confined nature of the aquifers and lack of primary porosity in rocks), there are limits to utilizable groundwater recharge.

This chapter focus on one such Indian state, ie, Maharashtra, which has mostly groundwater in hard rock formations and experiences acute drinking water scarcity during summer months. In this chapter, we compare the performance of selected rural water supply schemes (RWSSs) in the state and discover the techno-institutional model of water supply which works best under such settings. The chapter is divided into seven sections: Section 5.1 is the introduction; Section 5.2 provides the objectives and methodology adopted for the case study; Section 5.3 discusses the water supply reforms being undertaken in the state; Section 5.4 highlights the evolution of techno-institutional setup for water supply management in the state; Section 5.5 presents the comparative performance analysis of selected scheme types; Section 5.6 presents the overall findings; and Section 5.7 is the conclusion and includes policy inferences drawn from the study.

5.2 OBJECTIVES AND METHODOLOGY

The major objective of the study was to undertake a comparative performance assessment of different types of rural water supply schemes in order to understand the influence of policy reforms, techno-institutional characteristics, and water supply administration on scheme performance. The analysis covered the managerial (physical, economic, and financial) performance, governance, decentralization, and community participation.

For the purpose, a total of 12 schemes representing six different regions of Maharashtra were selected to analyze the scheme performance. Out of these, seven were individual piped water supply schemes and five were regional piped water supply schemes (Table 5.1). While sampling of the scheme, care was taken to ensure that two different types of schemes from the same type of geo-hydrological and hydrological setting were selected for comparison. This is to nullify the influence of the physical environment on the performance of schemes.

In each scheme-covered village chosen for the survey, a total of 50 households were surveyed to understand aspects such as water supply adequacy, reliability, and quality of water. From each chosen individual water supply scheme, one village, and from each chosen regional water supply scheme, two villages were selected for the survey. Thus, in total, 17 villages and 850 households were surveyed.

An understanding of the governance in rural water supply was developed through interview of the stakeholders, local political leaders, civil society groups, senior officials of state rural water supply bureaucracies, officials of regulatory agencies, and the user communities (consumers) on the effectiveness of the roles of every other stakeholder (either in rule making or rule implementing) on one or more of the following: planning water supply schemes; quality of water supply services for rural areas; fixing water prices or water taxes; taking investment decisions in water supply infrastructure; and operation and maintenance of rural water supply schemes. Since the focus is on understanding how far governance of rural water supply is decentralized, these observed practices were compared against those which promote decentralization.

The institutional and management performance of the utilities was studied by analyzing data collected from the managers of the schemes and communities which were served by them on various physical, socioeconomic, and institutional factors which determine these performance variables. The water administration was studied by analyzing the data from the state-level and local-level bureaucracies concerned with drinking water supply.

Table 5.1 Details of the selected schemes from the state of Maharashtra

Administrative division	District	Type of scheme	Block	Year of completion
Nagpur	Bhandara	Individual GP scheme based on dug well	Bhandara	2011
	Gondia	RWSS based on surface reservoir	Amgaon	2007
Amravati	Amravati	Individual GP scheme based on dug well	Nandgaon kh.	2001
	Yavatmal	RWSS based on surface reservoir	Pusad	2006
Aurangabad	Hingoli	Individual GP scheme based on open (infiltration) well	Hingoli	2010
	Osmanabad	RWSS based on surface reservoir	Osmanabad	2005
Nashik	Nandurbar	Individual GP scheme based on bore well	Shahade	2011
	Jalgaon	Individual GP scheme based on reservoir	Bhusawal	2011
Pune	Solapur	RWSS based on river lifting	Sangola	2005
	Pune	RWSS based on surface reservoir	Mawal	1997
Konkan	Thane	Individual GP scheme based on percolation well	Dahanu	2010
	Sindhudurg	Individual GP scheme based on river lifting	Sawantwadi	2005

Comparison between two different techno-institutional models of water supply from similar physical setting within a state was carried out to understand the influence of techno-institutional characteristics on the management performance of schemes. Such analysis was done under two geological settings. Comparison between schemes with different organizational set ups and operational policies from within the state was undertaken to analyze the influence of water supply administration on scheme performance.

5.3 RURAL WATER SUPPLY REFORMS IN MAHARASHTRA

According to the Census of India (GoI, 2011), Maharashtra has nearly 23.83 million households, with 13 million households in rural areas and 10.8 million households in urban areas. About 68% of these have access to tap water sources. The corresponding figure in rural areas is about 50%. Obviously, the progress in the rural areas is far from being satisfactory. But still, these figures show an improvement from 2001, when only about 45.5% of the rural households had access to tap water. Furthermore, there is some improvement in the proportion of rural households with drinking water available within their premises (increased from 38.9% in 2001 to 42.9% in 2011).

In Maharashtra, a demand-driven approach for delivery of rural water supply services was made operational in 2000 (Bassi et al., 2014; Prasad et al., 2014). Prior to that, during the early 1980s, there was an emphasis on single village schemes which mostly tapped groundwater. Regional rural water supply schemes started in a big way in the late 1980s, and particularly, during the mid- to late 1990s. These schemes drew water from large reservoirs, and were considered to be better than the single village schemes based on underground water as they involved economies of scale and the water was of better quality (Sangameswaran, 2010). Between the mid-1980s and mid-1990s, the state government implemented various schemes for improving the water supply coverage through financial support from the World Bank, the German Development Bank, and DFID. In 1995, Maharashtra became the first state in the country to prepare a White Paper on the water situation in the state (Prasad et al., 2014) and to initiate institutional reforms with a view to improving the performance of local bodies that are responsible for the provision of drinking water and sanitation facilities (Das, 2006). The White Paper primarily addressed the issues of disparity in water supply, level of service, and groundwater depletion. A few recommendations were made concerning water supply in rural areas, enforcement of various legislations, stressing upon the more effective management of water supplies and decentralization.

For drinking water, a master plan consisting of a number of detailed proposals was adopted in 1996. The overall goal of the proposals was to make Maharashtra free of tankers—a major source of drinking water for villagers during summer months—by the year 2000. As per the master plan, water supply norm became more stringent, with the supply rate increasing from 40 to 55 L per capita per day (lpcd). More importantly, a source dependability criterion of 95% was adopted for water sources. This meant that the surface water schemes had to be taken up in a big way, as groundwater-based schemes were found to be less dependable, with sources drying up during the summer months. Thus, the regional supply schemes acquired prominence. This led to the establishment of a parastatal agency called the Maharashtra Jeevan Pradhikaran (MJP) (Sangameswaran, 2010). The MJP built, tested, and then handed over schemes to the concerned local authorities to operate and maintain. For each completed scheme, MJP charged a commission of 17.5% on the total costs to meet its overhead costs and other administrative expenses. However, the tanker-free program was not a complete success. It required huge financial outlay; and the regional supply schemes brought in their own problems which included high capital costs, lack of willingness of local authorities to take over the schemes, and inequity in distribution of water between head-end and tail-end villages (Sangameswaran, 2010).

With the change in the state government in 1999, the centrally sponsored "Sector Reform Programme" was initiated in four districts of the state. By the year 2000, in principle, the role of government was shifted from service provider to that of policy formulation and capacity-building agency. During the reform process, communities were encouraged to decide the water supply schemes of their choice and efforts were made to ensure their sustained involvement through cost sharing. Decisions on the type of scheme, the implementing body, ie, whether it should be the Village Panchayat (VP)[2], the District Panchayat (DP)[3] or a non-governmental organization (NGO), as well as the technical service provider (whether it should be MJP, the DP, or a private party) rested with the village assembly (VA) at least nominally. These ambitious reforms were in sharp contrast to the pre-2000 situation, where both the need for drinking water and the manner of

[2] Village Panchayat is a local self-government institution at the village level. Members of the village panchayat are elected by the village assembly which includes all the adult citizens of the village.

[3] District Panchayat is an apex organization of the three-tier structure of the local self-government institutions in India. The other two are Block Panchayat (at the block level) and Village Panchayat (at the village level).

its provision was determined by state departments, parastatal agencies such as MJP, DPs, and Block Panchayat.

Furthermore, a major role was to be played by village water and sanitation committees (VWSCs), which are elected by the VA and are technically subcommittees of the VP. They would have funds directly devolved to them, would maintain accounts separately (from the general accounts of the VP) and plan and implement the scheme autonomously (Sangameswaran, 2010). Again, during this phase there was a technological shift from regional supply schemes based on reservoirs to single-village schemes based on groundwater. It was believed that mobilizing local community and costs recovery, which are important components of the supposedly "demand-driven approach," would be easier in small single-village schemes.

In 2002, German KfW funded a "demand-driven" scheme, called *Aaple Pani*, which was initiated in three districts of the state. In 2003, the World Bank financed Maharashtra Rural Water Supply and Sanitation Project, also called Jalswarajya, was initiated to cover rural areas of the remaining 26 districts of the state. The overall objectives of Jalswarajya were to increase access to rural water supply and sanitation services and to institutionalize their decentralized delivery by local governments (RSPMU, 2004). Jalswarajya relied on voluntary participation by communities, wherein communities select water supply and sanitation services from a menu of options, and targeted the provision of these services by project administrators (Pattanayak et al., 2007). As per the project guidelines, community contribution to the capital cost was kept at 5% for tribal villages and 10% for non-tribal villages; and communities were also required to bear 100% of the operation and maintenance (O&M) costs. The rest of the capital cost was met by the project (World Bank, 2011).

5.4 EVOLUTION OF TECHNO-INSTITUTIONAL SETUP FOR WATER SUPPLY MANAGEMENT

About 94% of the total geographical area of Maharashtra is underlain by hard rock formations. Because of the basic characteristics of the rock formations, physiography and interannual variability in the rainfall, there are severe limits on the occurrence of groundwater (GoM, 1999). However, traditionally, the majority of the villages in the state depend on wells for meeting their drinking water needs.

Until 1985, most of the rural water supply schemes in Maharashtra were based on either dug wells or bore wells. During 1985 to 1997, there was an increase in the number of piped water supply schemes which were based on surface water sources. During 1991 to 1998, some 17 single-village schemes based on groundwater and 50 regional schemes mostly based on surface water sources were implemented under the various externally supported projects (Das, 2006). The shift in water supply technology from groundwater sources to surface-water-based sources during this period was mainly due to: growing overexploitation and depletion of groundwater, which continued to threaten the physical sustainability of water supply sources; the general perception that regional schemes bring economies of scale, and quality of water from surface sources is better; and stringent water supply norms. These changes led to the emergence of a new kind of institutional setup, where MJP played an all-important role in constructing and handing over schemes to the concerned local authorities to operate and maintain. However, regional water supply schemes became unpopular, in spite of the fact that these reservoir-dependent schemes were more sustainable than groundwater-based schemes in rural areas in lieu of the fact that not much control could be exercised on groundwater withdrawal by farmers.

Among the many problems facing the regional water supply schemes, which are otherwise superior to groundwater-based single-village schemes, the lack of willingness on the part of local authorities to take over management is the most crucial. This lack of willingness mainly stems from the fact that the local authority, which in this case is the VP, lacks technical know-how, finance, management, and governance capabilities to run the schemes, which are sophisticated socio-technical systems, at the local level. Particularly, the fund availability for operation and maintenance is a major issue.

As a recent evaluation of Water and Sanitation Sector in India pointed out, decentralization is also fraught with problems at the local level due to issues of "accountability." In the case of Maharashtra, MJP takes up execution of rural water supply scheme on requests from concerned DP/VP and/or when the rural scheme cost is equivalent to or above US$ 833,000. On completion of such schemes, MJP is expected to hand over the scheme to DP (in the case of a multivillage scheme) and VP (in the case of small schemes) for operation and maintenance. In such cases, it receives the funds directly from the state water supply and sanitation department (World Bank, 2012). Hence, they have no special incentive to make sure that the system so built runs reliably and efficiently. It is no surprise that the VPs are typically reluctant to take up the operation and maintenance of the schemes.

As a result of these developments, village-level schemes (mostly based on groundwater) of water provision were again started to be looked upon as a better alternative. Along with this shift in the scale of organization of water systems also came the shift from a supply-driven approach to a demand-driven approach (Sangameswaran, 2010). Under this new approach, efforts were made to empower local-level institutions to demand water supply schemes from the governmental authorities. Furthermore, VA and VWSC were given the authority to decide on the kind of scheme, and also to implement, operate, and maintain them. The official authority of the state government departments (including MJP, DP) was considerably reduced. However, this demand-driven approach largely remained only on paper. Without technical capacity to scrutinize the plan and design of the schemes, and techno-managerial capabilities to run them, the local Panchayats and VWSCs were not able to exercise the powers vested with them.

This is evident from the fact that out of 13,515 completed schemes, only 1951 schemes were handed over to VP. Most of the success in terms of community contribution to the capital cost of the water supply infrastructure came only in the schemes which were externally aided and had a huge financial allocation for the community mobilization. Furthermore, the inability of the VWSC to encourage users to pay a water fee affected the overall financial recovery, which also adversely impacted on the overall performance of the schemes. Till March 2007, the dues outstanding in respect of execution and maintenance of rural piped water supply schemes from ZP/VP amounted to about US$ 84 million. Dues outstanding with respect to maintenance and repair work undertaken by MJP are major and those related to community (popular) contribution are only minor (Fig. 5.1). Moreover, as almost 80% of the implemented schemes during the reform era tapped groundwater, the

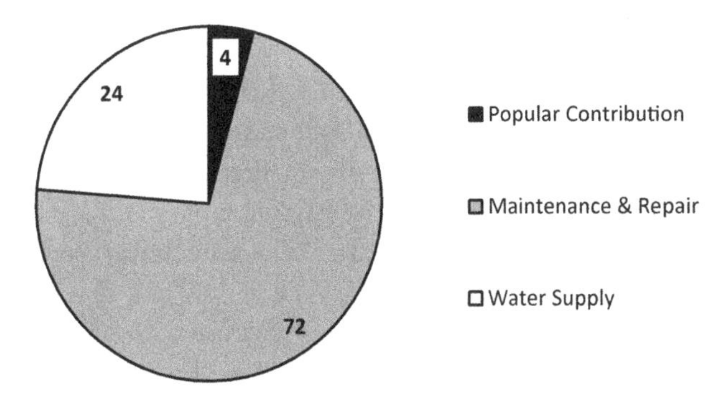

Figure 5.1 Contribution of various particulars to total outstanding dues from DP/VP till March 2007. *MJP Annual Report 2007–08.*

ability of such sources to supply water in summer or during drought years was questionable.

5.5 PERFORMANCE OF RURAL WATER SUPPLY SCHEMES IN MAHARASHTRA: A COMPARATIVE ANALYSIS

5.5.1 Institutional Structure for Managing Rural Water Supplies

Water Supply and Sanitation Department (WSSD), GoM is the state nodal agency for formulating, implementing, operating and maintaining regional water supply schemes in both rural and urban areas. The Groundwater Surveys and Development Agency (GSDA), the MJP, and Water and Sanitation Support Organization (WSSO) are the three line agencies supporting the WSSD. The GSDA is a technical agency (mostly geologists), and is entrusted with the responsibility of overall development and management of groundwater. Its Directorate is located in Pune, which is assisted by six regional and 33 district level offices. The MJP mainly consists of engineers and implements the piped water supply schemes. The MJP, with central offices in Mumbai and Navi Mumbai, has field offices spread across the entire state. Overall, there are five zonal offices, 16 circle offices, 44 work/project divisions, and 151 subdivisions.

As per the changes brought about by the Government Resolution (GR) of June 2003, some of the functions and functionaries of the GSDA and MJP were transferred to DP. The Water Supply Department of DP mainly comprises these transferred functionaries and is responsible for implementing water supply and sanitation reform programs. A Reform Support and Project Management Unit (RSPMU) was also set up in order to facilitate the rural water supply reforms process. The RSPMU operates at the state and district level.

WSSO was established to implement and monitor the state-level and national rural water supply program in Maharashtra as per guidelines of the central government. The main objective of this organization is planning, implementation and monitoring of rural water supply and sanitation projects and programs sponsored by central and state government and also financed by external agencies.

Furthermore, at the local level, DP are entrusted to provide technical service and approve funds to villages served by rural water supply schemes, whereas VP is responsible for demanding new drinking water supply schemes from the DP. VA is also empowered to decide on the kind of scheme, the implementing body, as well as about the technical service provider. The lowest in the tier is the VWSC which plans, implements, operates and maintains the village water supply scheme autonomously.

Once rural water supply schemes are planned, they are taken up for implementation with administrative approval and technical sanction of the designated authorities. For a single-village piped scheme up to US$ 833,000, VPs have been given responsibility for execution with the technical support from DP and MJP. Whereas for the single-village schemes costing above US$ 833,000, MJP is in charge of execution and the role ofVP is confined to operation and maintenance.

For multivillage schemes costing up to US$ 417,000, DP is responsible for the entire execution and role of VP is confined to O&M and that too under the technical supervision of DP. For the multivillage schemes above US$ 417,000, responsibility of execution shifts to MJP. Here again the role ofVP is restricted to O&M of the system. Thus, it is quite clear that the role ofVP/VWSC is mainly restricted to O&M of the schemes. For the project with high cost, even the O&M function is performed under the technical supervision of DP or MJP.

5.5.2 Physical Performance

5.5.2.1 Scheme Coverage

As regards coverage, 45% of the rural households were found to be covered by formal water supply in the selected schemes. There are significant differences in coverage across water supply technologies (Fig. 5.2). The highest

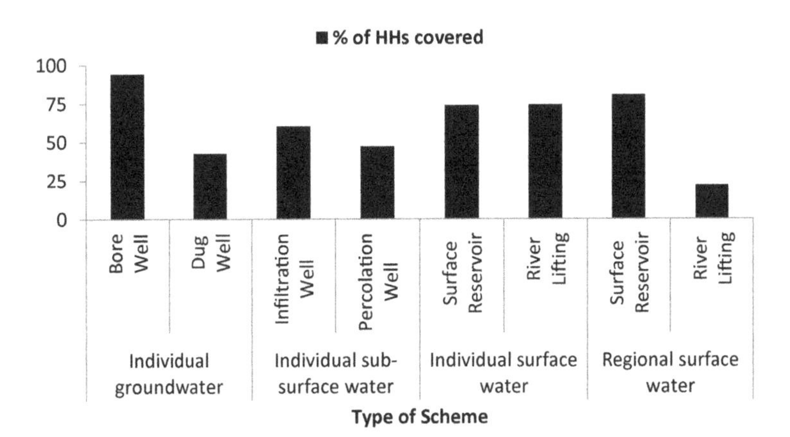

Figure 5.2 Household covered under different water supply schemes. *Authors' own analysis using primary data collected from various water supply agencies.*

coverage was seen to be achieved in a single-village scheme based on bore well, followed by multivillage schemes based on surface reservoir, and single-village schemes based on river lifting and reservoir. The regional scheme based on river lifting had the lowest coverage. The single-village schemes based on dug wells, percolation wells, and infiltration wells also have relatively lower percentage coverage as compared to reservoir-based single- and multivillage schemes. Notably, there is a remarkable difference in coverage between schemes that tap surface reservoirs and those that lift river water directly, when the scheme has to supply water for a large number of villages (74% against 22%). Differences in coverage were also seen with change in the type of source. In the case of single-village schemes, those tapping water from surface sources (74%) were found to be performing better than those tapping groundwater (64%) and subsurface water (57%).

As regards the proportion of households with individual tap connections, it was highest for single-village schemes based on river lifting and surface reservoir (about 74% each) and the lowest (35.6%) for single-village schemes based on subsurface water. Nearly 63% of rural households that were covered by a regional water supply scheme based on reservoirs also had individual household tap connections. Differences in coverage were also seen with change in type of source. In the case of single-village schemes, those tapping water from surface sources were found to have the highest percentage of individual tap connections (74%), followed by those tapping groundwater sources (48%) and those tapping subsurface water had the lowest percentage of households with an individual tap connection (36%).

Overall, it appears that the water supply schemes based on surface reservoirs, both single-village schemes and multivillage schemes, show relatively better performance as compared to single-village schemes tapping groundwater and subsurface water, and regional schemes lifting water directly from the river, in terms of the proportion of the households actually served by them and proportion of households provided with individual tap connections.

5.5.2.2 Adequacy

In terms of daily per capita supply, all the schemes were found to be supplying at par with or more than the state-adopted norm of 40 lpcd in all seasons, except in the case of the one individual schemes (tap subsurface flow using infiltration well) that supplied only 32 lpcd during summer months. The estimated per capita supply levels maintained by different types of schemes studied are presented in Table 5.2. The table shows that overall the

Table 5.2 Supply levels maintained by different types of selected scheme

Type of scheme	Source	Average amount of water to be supplied as per design (lpcd)	Actual amount of water supplied per scheme (lpcd)		
			Monsoon	Winter	Summer
Individual ground water	Bore well	40	40	40	40
	Dug well	47.5	117.5	117.5	110
Individual subsurface water	Infiltration well	–	40	40	32
	Percolation well	40	80	80	80
Individual surface water	Surface reservoir	40	40	40	40
	River lifting	40	70	40	40
Regional surface water	Surface reservoir	55	46	49	46
	River lifting	55	55	55	55

Authors' own analysis using primary data.

average per capita water supply is slightly higher for groundwater-based schemes than their counterparts based on reservoirs and river lifting. However, this is because in dug-well-based schemes, an excessively high level of supply is achieved primarily due to a small percentage of the targeted households (only 42.7%) being served.

In the overall assessment what emerges is that in the case of single-village schemes based on groundwater, percolation wells, and infiltration wells, the average level of water supply per capita is slightly better due to the extent of coverage of the targeted households. For instance, the schemes that tap surface reservoirs, supply water to 73.7% of the total households considered in the design. Whereas the schemes that extract groundwater and subsurface water supply water to a significantly lower proportion of households (57%). Another important factor which needs to be considered while analyzing the performance of the schemes is that the norms for per capita water supply considered for designing the scheme is in the range of 40–55 lpcd. While the targeted water supply is achieved in the case of single-village, reservoir-based schemes, the actual per capita supply is slightly less than the norm (of 55 lpcd) for the regional water supply schemes.

Within the scheme, no major variation is seen in the frequency of water supply across different seasons. Amongst schemes, the frequency of water supply ranged from twice a day to once in 2 days for single-village scheme based on groundwater; twice a day for a single-village scheme based on surface reservoir; once a day for an individual village scheme based on river lift; once a day to once in 2 days for regional schemes based on both surface reservoir and river lift. In terms of hours of daily water supply, regional schemes based on surface water fare much better than any other schemes (Table 5.3). On average, such schemes were found to supply water for 7.9 h per day during monsoon; 5.6 h per day during winter; and 6.1 h per day during summer. Individual schemes that abstract groundwater using bore wells showed the poorest performance, supplying water only for 1 h during monsoon and winter; and for 0.67 h per day during summers. As discussed earlier, the sustainability of schemes based on groundwater in supplying water during summer months is uncertain.

A realistic assessment of the comparative performance of the schemes cannot be made merely on the basis of data on per capita supply levels maintained by the schemes. It is also important to know the current per capita

Table 5.3 Frequency of water supply in different types of water supply schemes

Type of scheme	Source	Average hours for which water to be supplied as per design (hour per day)	Average hours of water supply per day		
			Monsoon	Winter	Summer
Individual ground water	Bore well	8	1	1	0.7
	Dug well	4.5	1.5	1.5	1.5
Individual subsurface water	Infiltration well	–	1	1	1
	Percolation well	8	4	4	4
Individual surface water	Surface reservoir	1.5	1	1	1
	River lifting	–	1	1	1
Regional surface water	Surface reservoir	10	3.8	4.2	4.8
	River lifting	20	20	10	10

Authors' own analysis using primary data.

requirements as this can change from region to region and from location to location depending on the socioeconomic and climatic conditions and cultural settings. Once that is known, one should know as to what extent the requirements are met from the new scheme and the per capita water use. The latter indicators are perhaps more important as many villages have dual schemes, an old one and the new one (which includes common wells, hand pumps, etc.) and people might be able to tap water from them as well.

The analysis of data on household-level water use shows that per capita water use is highest for households under a single-village river lift scheme (around 80 lpcd), followed by 75 lpcd for dug-well-based schemes and 59 lpcd for regional schemes based on reservoirs. The lowest per capita water use was found in the case of bore-well-based schemes (28.6 lpcd). What is interesting is that the per capita requirement was also proportionally lower for bore-well-based schemes, and that the entire requirement is met by the scheme. One possibility is that due to water shortage, the communities are reducing their water demands by not going for economic activities such as livestock keeping. In other cases, where the water use is high, the communities might have demanded a new scheme to meet their domestic water requirements like in the case of the dug-well-based scheme, or the single-village reservoir scheme or the multivillage scheme.

As regards the extent of fulfillment of the household water requirement, there is significant improvement with the introduction of new schemes in all except the case of a multivillage scheme based on reservoir (Fig. 5.3). The largest change is in the case of an individual scheme tapping subsurface

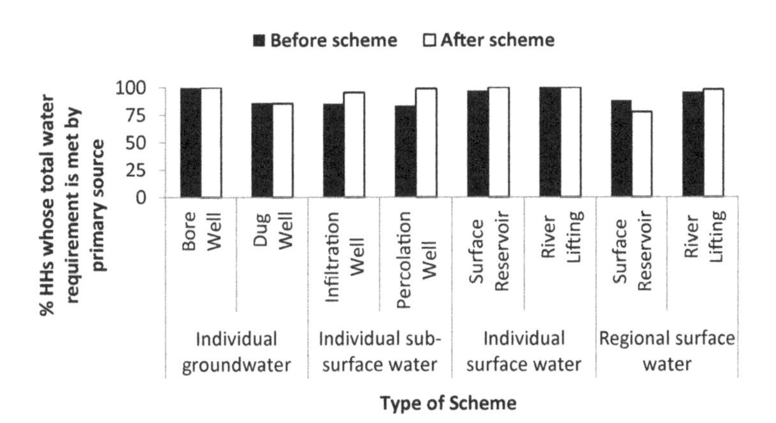

Figure 5.3 Extent to which water requirements are met by the primary source pre and post new scheme. *Authors' own analysis using primary data.*

water. Marginal improvement is also observed for the bore-well-based schemes, surface reservoir-based single-village schemes, and river-lifting-based multivillage schemes. However, in the case of the multivillage scheme based on reservoir, the situation deteriorated. This might have happened because the coverage of the scheme in terms of number of villages and households (HHs) might have increased, and many of those which were earlier not covered by PWS are covered by the newly introduced scheme, and the new source functions as a supplementing source. This is partly explained by the fact that nearly 45% of the households depend on secondary as well as primary sources to meet their household water needs. Nevertheless, in the case of single-village schemes based on surface water, currently 100% of the requirement is met from the new source, as compared to 97.3% for subsurface schemes and 90.4% for schemes based on groundwater.

5.5.2.3 Access and its Link to Quality of Water Supply

The degree of access to the source and the percentage of households (HHs) which exclusively depend on the primary source can be good indicators of the quality of water supply from a source. The first one would decide whether the households are able to access water from a particular source with ease or not. Here, we would consider only those who access water from the source within the premise as well as those who have an individual HH tap connection. The second would indicate how much importance the source has in meeting the water needs of the households covered by the scheme. The outputs of the analysis carried out using the data collected from the field in this regard are presented in Table 5.4.

Overall, amongst different water sources, the degree of access to water supply source appears to be much higher for single-village schemes based on surface water (86%) as compared to those based on groundwater (62.3%) and subsurface sources (35.3%). The figures however change with change in technology. In the case of bore-well-based schemes the degree of access is the highest (92%), followed by single-village schemes based on surface reservoir (86%) and river lift (84%). It is 71% for infiltration well-based schemes and 54% for multivillage schemes based on surface reservoirs. But, it is also noteworthy that in the case of bore-well-based schemes, the average amount of water used by the households is much less.

The proportion of households which depend exclusively on the primary source is higher for surface-water-based schemes. Across technologies, in the case of bore well schemes and single-village schemes based on river lift, all the households depend exclusively on the primary source.

Table 5.4 Access to water supply from the sample sources and degree of dependence

		Proportion (%) of total households with				
Type of scheme	**Source/ Technology**	**Individual HH tap connection**	**Tap connection within the dwelling**	**Stand post near dwelling**	**HHs who exclusively depend on the primary water source (%)**	**HHs depend on both primary and common sources (%)**
Individual groundwater	Bore well	40	52	8	100	0
	Dug well	47	0	0	47	53
Individual subsurface water	Infiltration well	71	0	0	71	29
	Percolation well	0	0	0	0	100
Individual surface water	Surface reservoir	51	35	12	98	2
	River lift	66	18	16	100	0
Regional surface water	Surface reservoir	36	18	0	54	46
	River lift	50	0	0	50	50

Authors' own analysis using primary data.

On the other hand, none of the households covered by the scheme supplying water through a percolation well depend exclusively on the same. A significantly high proportion of selected households under the surface-reservoir-based single-village scheme depends exclusively on the primary source. In the case of multivillage reservoir-based schemes, more than half of the total households depends exclusively on the primary source, and the remainder depend on both the primary and common sources. The importance of the source is also highest for single-village schemes based on surface water, with 98% of households exclusively depending on the primary source.

5.5.2.4 Distance to the Source

One other important indicator of measuring the performance of a water supply scheme is the average reduction in distance to the source (Cairncross and Kinnear, 1992; Howard, 2002; Alegre et al., 2006; Hoko and Hertle, 2006; Bassi et al., 2014). As a result of introduction of an improved source of water, the distance traveled by the members of the household to collect water has reduced considerably after the introduction of new schemes. The maximum reduction in distance traveled per sample households was found in the surface-reservoir-based multivillage scheme, which is followed by the surface-reservoir-based single-village scheme; subsurface-water-based single-village scheme; bore-well-based single-village scheme. The least reduction was in the case of the regional river lift schemes (Fig. 5.4).

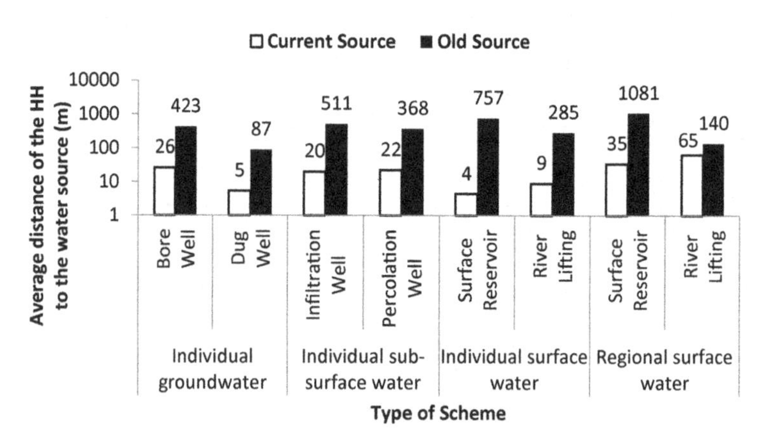

Figure 5.4 Scheme-wise average distance traveled by households to access water from primary source. *Authors' own analysis using primary data.*

5.5.3 Economic Performance: Cost of Water Supply

Capital cost, in terms of rupees per unit volume of water supplied, was found to be highest (INR 0.6 per liter) for regional schemes that are based on surface reservoirs. Lowest was for individual schemes that are based on surface water (INR 0.1 per liter). Among the schemes that are based on groundwater and subsurface water, the scheme based on percolation well has the highest capital cost (INR 0.4 per liter) of water supplied, which is followed by the one based on infiltration and bore wells (INR 0.2 per liter each); and dug wells (INR 0.1 per liter). But, one needs to use these figures with caution. There are many reasons for this. First: the life of the system would differ drastically with change in scheme type. Though the water distribution infrastructure would have the same life across schemes, the life of the source would change across resource types. A large or medium reservoir which taps surface water from a large catchment would generally have a long life. At the same time, a well, which taps aquifer in the hard rock area, would have a very short life of 10–12 years. Second: wells are not substitutes for regional water supply schemes. Instead, regional water supply schemes are resorted to when local groundwater-based sources fail due to droughts or resource depletion. Since they address the specific problem of source sustainability, comparison of costs between such schemes would be inappropriate.

As regards the average O&M cost per household for the HHs covered by formal water supply, the highest was for the regional schemes based on river lifting, followed by single-village schemes based on river lifting (Table 5.5). One reason for the exceptionally high O&M cost for the regional water supply schemes based on river lifting is that a relatively small proportion of the households considered for design of the scheme actually benefit from it. This could be due to poor yield from the catchment. The average operational cost per household was lowest for the bore-well-based scheme and schemes based on infiltration well and percolation well. Interestingly, for the reservoir-based multivillage scheme, the average annual O&M cost per HH was slightly lower than the O&M cost for the dug-well-based scheme. This is contrary to the general perception that regional water supply schemes incur high O&M costs. The O&M cost was much lower for the single-village scheme based on surface reservoir which is less than half that of the dug-well-based scheme.

However, to understand what factors contribute to the cost differences, it is necessary to look at the break up of these cost figures. The output shows that the proportion of average maintenance and repair (M&R) cost/HH to the total O&M cost/HH was lowest for the regional scheme lifting river

Table 5.5 Annual O&M cost per HH for different types of water supply schemes

Type of scheme	Source	Average cost of construction (INR Million)	Average annual O&M cost per scheme (Rs/HH/)	Contribution to average annual O&M cost (%)			
				Salaries	Electricity Charges	M&R	Others
Individual ground water	Bore well	2	228	24	15	59	1
	Dug well	3	1181	45	23	24	7
Individual subsurface water	Infiltration well	5	323	40	24	20	15
	Percolation well	33	369	39	43	14	4
Individual surface water	Surface reservoir	7	445	29	39	19	13
	River lifting	2	1190	31	40	29	0
Regional surface water	Surface reservoir	160	1139	34	46	12	8
	River lifting	991	5468	15	67	2	16

Authors' own analysis using data furnished by MJP.

water, followed by the surface-reservoir-based regional scheme and the percolation-well-based scheme. It was significantly higher for the individual village scheme based on bore well. Contribution of salaries to the total O&M cost/HH was quite high in the case of dug-well-based individual schemes, next only to infiltration-well-based individual water supply schemes. Further, average electricity charges were significantly higher for regional schemes based on river lifting (INR 1155/HH), as water transport and distribution involved many stages of pumping to take water to high elevations. The electricity charges as a percentage of the annual O&M cost varied from 15% for single-village bore-well-based schemes to the highest of 67% for a regional water supply scheme based on river lifting. It is seen that the electricity charges significantly contributed to the differences in O&M cost across scheme types.

5.5.4 Financial Performance: Revenue Recovery

Average water tariff per household connection varied from INR 30/month/connection in bore-well-based groundwater schemes to INR 75/month/connection in surface-reservoir-based regional schemes. For other covered households under regional water supply schemes, average bulk tariff varied from INR six to eight per kiloliter. Overall, average tariff collection (tariff recovery as a percentage of the total demand), was significantly higher for schemes that tap surface water (70.5%) than those which tap groundwater and subsurface water (63%). It is to be kept in mind that the surface-water-based schemes also have a significantly high proportion of the households under individual tap connections (Table 5.6). As regards technology, average

Table 5.6 Revenue recovery in the selected water supply schemes

Type of scheme	Source	Average tariff collection (%)	HHs with individual piped connection (%)
Individual ground water	Bore well	79	56
	Dug well	80	43
Individual subsurface water	Infiltration well	59	60
	Percolation well	18	24
Individual surface water	Surface reservoir	70	74
	River lifting	72	74
Regional surface water	Surface reservoir	63	63
	River lifting	100	0

Based on Bassi, N., Kumar, M.D., Niranjan, V., Kishan, K.S.R., 2014. The decade of sector reforms of rural water supply in Maharashtra. In: Kumar, M.D., Bassi, N., Narayanamoorthy, A. Sivamohan, M.V.K. (Eds.), The Water, Energy and Food Security Nexus: Lessons from India for Development, Routledge, UK, pp. 172–196.

tariff collection was highest for the regional scheme lifting river water, which is followed by schemes extracting groundwater through dug wells and bore wells; individual schemes tapping water through river lift and surface reservoir; regional schemes dependent on surface reservoir; and schemes extracting subsurface water through water through infiltration and a percolation well.

5.5.5 Decentralization and Community Participation

Under the sector reforms, the Village Water and Sanitation Committee is expected to perform many functions relating to planning, design, execution, and management of water supply at the local level for achieving decentralized management and community participation. For understanding the degree of decentralization in management, observed roles and responsibilities of VP/VWSC were compared with those that are expected.

As regards planning and designing of schemes, in 10 out of 12 selected schemes, planning and design of water supply schemes was undertaken by either the technical wing of DP or MJP. Only in two individual schemes, one based on dug well and another on percolation well, planning and design of the scheme was carried out by VWSC. Further, in only 50% of the selected cases, VP built the scheme but they were found to operate all the selected individual schemes. However, as per the approved administrative procedures, all single-village piped schemes with project cost ranging from US$ 83,000 to US$ 833,000 had to be executed by VP. But MJP and DP were found to have executed two of the selected seven individual schemes (Table 5.7).

In the case of single-village schemes, in most cases, the scheme was built by the VP, while in one case, the scheme was built by DP and in another the scheme was built by the MJP. Nevertheless, in all cases, the scheme was operated by the VP. In the case of regional schemes, MJP emerged as the major player performing the function of main system management, while the operation and maintenance of the scheme components at the village level is done by the VWSC. However, that too is done under the technical supervision of MJP. It is quite clear from this arrangement that the majority of management functions are in the hands of agencies such as MJP, and the DP or the VP do not take charge of the system operation.

Table 5.8 provides the details of the expected roles and number of VWSC actually performing these roles. In the case of single-village schemes, only a few functions are performed by every VWSC. Seven out of the 13 identified functions are not performed by some of the VWSCs. A marked difference is found between the VWSC of regional schemes and that of single-village

Table 5.7 Details about decision on various aspects of water supply in selected schemes

Type of scheme	Technology	Scheme planned and designed by	Scheme built by	Scheme operated by	Institution managing WSS in village
Individual ground water	Bore well	DP	VP	VP	VP
	Dug well	VWSC, MJP	VP, MJP	VP	VP, VWSC
Individual subsurface water	Infiltration well	DP	DP	VP	VWSC
	Percolation well	VWSC	VP	VP	VWSC
Individual surface water	Surface reservoir	DP	VP	VP	VWSC
	River lifting	DP	VP	VP	VWSC
Regional surface water	Surface reservoir	MJP	MJP	MJP, VWSC	MJP, VWSC
	River lifting	MJP	MJP	MJP	VWSC

Authors' own analysis using primary data collected from various water supply agencies.

Table 5.8 Roles performed by the VWSC against the expected functions

		% of VWSC performing the following functions	
Sr. No.	**Functions expected of VWSC**	**Regional schemes (total of 183 VWSCs)**	**Individual schemes (total of 7 VWSCs)**
1.	Ensuring community participation and decision-making in all phases of scheme activities	31	100
2.	Organizing community contributions towards capital costs	0	100
3.	Operating bank account for depositing community cash contributions and O&M funds	31	86
4.	Preparation of village water security plan	0	43
5.	Planning, designing, and implementing all water-related activities in the village	31	100
6.	Planning, designing, and implementing all sanitation-related activities in the village	31	43
7.	Procuring construction materials/goods and selection of contractors (where necessary) and supervision of construction activities	31	86
8.	Ascertaining drinking water adequacy at the household level including cattle needs	26	57
9.	Tariff collection	31	100
10.	Empowering of women for day-to-day operation and repairs of the scheme	0	71
11.	Participation in communication and development activities in other villages	5	29
12.	Testing of supplied water quality	31	100
13.	O&M supervision and monitoring	31	100

Based on Bassi, N., Kumar, M.D., Niranjan, V., Kishan, K.S.R., 2014. The decade of sector reforms of rural water supply in Maharashtra. In: Kumar, M.D., Bassi, N., Narayanamoorthy, A. Sivamohan, M.V.K. (Eds.), The Water, Energy and Food Security Nexus: Lessons from India for Development, Routledge, UK, pp. 172–196.

schemes in terms of the number of functions being performed by the VWSC. In the case of regional water supply schemes, less than one-third of the village water supply and sanitation committees perform some of the roles vested in them (eight out of the 13). In other cases, the committees are largely defunct. Certain activities are not performed by any of the VWSCs. For instance, none of the VWSC of the regional schemes take up roles like preparation of village water safety and security plan. Hence, it can be summarized that overall involvement of the village communities in the management of rural water supply is less for regional schemes as compared to local schemes.

5.5.6 Water Supply Governance

Governance is the art of rule-making. In the context of water supply, this can pertain to planning of water supply schemes; quality of delivery of water supply services; fixing of water prices or water tax; and investment for water supply infrastructure (Kumar, 2014).

The results show that in the case of single-village schemes, the governance of rural water supply is more or less decentralized, with mostly the VP deciding on the type of scheme and source, water supply schedules, duration and timing of water supply, the individual connection charges, mode of pricing water and water charges, penalty for non-payment of water charges wherever it exists, and frequency of water quality monitoring. To an extent, the DP is also found to be involved in performing some of the governance-related functions. Contrary to this, in the case of a regional water supply scheme, the DP and MJP together replace the VP/DP, with some of the governance roles being performed only by the MJP. For instance, in the case of the regional scheme based on reservoirs, the MJP was perceived to have a significant role in deciding on the type of scheme, type of source, water supply schedule, timing and duration of water supply, individual tap connection charges, mode of pricing water and water rates, and frequency of water quality monitoring. At the same time, in the case of single-village schemes, most of these decisions are taken by the VP and to an extent by the DP.

5.5.7 Human Resource Capabilities

Though there are no standard norms being followed in the state regarding the number of agency staff required to handle water supply; distribution; and maintenance functions, it was found that the number of staff handling such functions was extremely low. Overall per 1000 households covered by water supply, only 0.33 technical staff; 0.21 managerial staff; 0.14 financial staff; 1.27 other non-technical staff; and 2.94 contract-based staff were found to be handling the water supply functions. The number of technical staff was too low considering that the major functions for executing projects in the case of individual village schemes and almost all the functions in the case of regional water supply schemes are handled by them.

Among the schemes, the highest number of technical staff per 1000 covered households was found in individual groundwater-based schemes (1.2) and lowest in individual surface-water-based schemes (no employee). Managerial staff was also highest in the case of groundwater-based schemes (3.5 per 1000 covered HHs) and lowest in individual surface-water-based schemes (no employee). However, non-technical staff per 1000 covered households was highest in individual surface-water-based schemes (4.3) and lowest in regional surface-water-based schemes (1.1). Further, only surface-water-based regional schemes had employed contractual staff.

5.6 MAJOR FINDINGS

Over the past 10 years, the state of Maharashtra has made significant progress in improving the access of rural communities to domestic water supplies, in terms of physical access to the sources. However, even today, only half the rural households have access to tap water, and this includes those who access it from distant sources.

A comparative assessment of the performance of rural water supply schemes in Maharashtra shows that the management performance of single-village schemes, which are based on surface water, is better than that of their groundwater and subsurface water counterparts. Also, the overall performance of water supply schemes based on surface reservoirs (both single village and multivillage) is better than that of schemes based on groundwater and subsurface water.

However, the degree of decentralization and community participation in management of the scheme is much better for the single-village schemes based on groundwater which were designed and built by the DP (in one case) and the VP (in four cases). In such schemes, all the VWSCs were reported to be performing only six out of the 13 key roles vested with them, while the remaining roles are performed by a lesser number of VWSCs. In the case of regional water supply schemes which were built and operated by the MJP, only in less than one-third of the cases (57 out of the 183 VWSC covered) were the VWSCs performing some of the key roles, while in the rest of the cases, they were totally dysfunctional.

Further, in the case of single-village schemes, most of the governance functions relating to water supply are performed by the local self-governing institution of the VP and some by the DP, whereas in the case of regional schemes based on reservoirs supplying water to many villages, MJP performs many of these functions along with the DP and VP.

Additionally, the single-village schemes based on surface water are run with a lower number of technical and managerial staff as compared to groundwater-based schemes in terms of number of staff per 1000 covered HHs, though they have a higher number of contractual staff as compared to groundwater-based single-village schemes. The regional schemes have the lowest number of contractual staff per 1000 connected households.

5.7 CONCLUSIONS AND POLICY INFERENCES

The reforms which started in the early 1990s have changed the way rural water supply is operated and managed by the state. New Acts, institutions, and policies were framed all through the course. Yet the overall coverage of villages and habitations with improved water sources remains low in rural areas. Furthermore, some of the landmark decisions taken earlier with regard to the choice of technology, i.e., shift of focus from groundwater-based individual supply schemes to surface-water-based regional supply schemes, were revoked and that had an adverse impact on the supply source sustainability.

As indicated by the performance assessment of 12 rural water supply schemes, heavy dependence on groundwater-based sources, on the premise that the local institutions such as the village Panchayats could run and manage these schemes, has led to poor system performance. However, the degree of decentralization and community participation in management of the scheme is comparatively better for the single-village schemes based on groundwater, this has come at the cost of sustainability of water supplies not only from the point of view of providing sufficient quantities of water for meeting domestic water requirements throughout the year, but also in terms of cost-effectiveness. Furthermore, because of the basic characteristics of the rock formations (95% of the state is underlain by hard rocks), physiography, and interannual variability in the rainfall, there are severe limits on the occurrence of groundwater in the state (GoM, 1999). Additionally, improper formulation and ineffective implementation of the groundwater Act has aggravated the problems as it has been largely ineffective in regulating groundwater use by irrigators for protection of drinking water sources.

Clearly, there is a tradeoff between "sustainability" and "decentralized governance and management." Even the goal of decentralization in governance and management of water supply scheme hasn't been achieved.

All the village water and sanitation committees were reported to be performing only six out of the 13 key roles vested with them, while the remaining roles were performed by a lesser number of VWSCs. The situation is even grimmer in the case of selected regional water supply schemes where only in less than one-third of the cases, were the VWSCs performing some of the key roles, while in the rest of the cases, they were totally dysfunctional.

Currently, the willingness on the part of the VP or DPs to take charge of running the regional water supply schemes is largely absent. Most of the regional schemes are still run by the MJP, while in a very few cases where the number of villages covered is small, the DP had taken over the running of the system. The important reasons for this are the technical sophistication of the schemes, the lack of qualified staff to take care of the maintenance, and fear of the financial burden of "high O&M costs." Furthermore, the response of the VWSC in terms of taking over the village-level maintenance of the scheme is also not encouraging as they are at best equipped to mobilize village community for social action. Therefore, handing over of responsibilities to them without building their technical, managerial and financial capabilities is leading to a situation of "leapfrogging."

In lieu of the water scarcity experienced in many districts of the state, the issues related to water supply quantity need to be tackled differently from those related to water quality. For that, the approach of planning, designing, and implementing rural water supply schemes as single-use systems (domestic use) has to be replaced by schemes which can take care of multiple needs (both domestic and productive use) of the village community. Furthermore, if sustainability of the source has to become a priority, then the strategy has to change from groundwater-based individual supply schemes to reservoir-based regional supply schemes, unless proper enforcement of groundwater regulation is done. With the large number of large and medium-sized reservoirs scattered across the state, and with around 2000 million cubic metre (MCM) of water remaining unutilized in these storage systems at the end of the irrigation season every year, augmenting water supplies of the existing village or regional schemes should be possible with the building of a large distribution network using pipelines (IRAP, 2012; Bassi et al., 2014).

On an institutional development front, the VWSC needs to be properly trained so that they can take effective part in design and O&M of multiple-use RWSSs. Thus, the community can feel the incentives and take active part in operation and maintenance of the systems. The scheme with multiple uses should consider all the available sources in the area in order to judiciously use the available resources. In the case of regional water supply schemes covering a large number of villages, the system operation should be handled by the MJP. As they are involved in planning and execution, such an approach would improve the operational efficiency. The financial and human resource capacities of the MJP need to be strengthened so that they could play an effective role in rural water supply management.

REFERENCES

Alegre, H., Hirner, W., Baptista, J.M., Parena, R., Cubillo, F., Cabrera Jr., E., Merkel, W., Duarte, P., 2006. Performance Indicators for Water Supply Services, second ed. International Water Association (IWA) Publishing, London, UK.

Bassi, N., Kumar, M.D., Niranjan, V., Kishan, K.S.R., 2014. The decade of sector reforms of rural water supply in Maharashtra. In: Kumar, M.D., Bassi, N., Narayanamoorthy, A., Sivamohan, M.V.K. (Eds.), The Water, Energy and Food Security Nexus: Lessons from India for Development. Routledge, UK, pp. 172–196.

Cairncross, S., Kinnear, J., 1992. Elasticity of demand for water in Khartoum, Sudan. Social Science Medicine 34 (2), 183–189.

Das, K., 2006. Drinking Water and Sanitation in Rural Maharashtra: A Review of Policy Initiatives. Gujarat Institute of Development Research, Ahmedabad.

Government of India (GoI), 2011. Census of India. Government of India, New Delhi.

Government of Maharashtra (GoM), 1999. Maharashtra Water and Irrigation Commission Report. Government of Maharashtra, Maharashtra.

Hoko, Z., Hertle, J., 2006. An evaluation of the sustainability of a rural water rehabilitation project in Zimbabwe. Physics and Chemistry of the Earth, Parts A/B/C 31 (15), 699–706.

Howard, G. (Ed.), 2002. Water Supply Surveillance: A Reference Manual. Water, Engineering and Development Centre (WEDC), Loughborough University, UK.

Institute for Resource Analysis and Policy (IRAP), 2012. Promoting Sustainable Water Supply and Sanitation in Rural Maharashtra: Institutional and Policy Regimes. Institute for Resource Analysis and Policy, Hyderabad.

Kumar, M.D., 2014. Thirsty Cities: How Indian Cities Can Manage Their Water Needs. Oxford University Press, New Delhi.

Pattanayak, S.K., Poulos, C., Wendland, K.M., Patil, S.R., Yang, J.C., Kwok, R.K., Corey, C.G., 2007. Informing the Water and Sanitation Sector Policy: Case Study of an Impact Evaluation Study of Water Supply, Sanitation and Hygiene Interventions in Rural Maharashtra, India. Working Paper 06_04. Research Triangle Institute International, NC, USA.

Prasad, P., Mishra, V., Sohoni, M., 2014. Reforming rural drinking water schemes: the case of Raigad district in Maharashtra. Economic and Political Weekly 49 (19), 58–67.

Reform Support and Project Management Unit (RSPMU), 2004. Maharashtra Rural Water Supply and Sanitation Project – Jalswarajya Quarterly Progress Report, January-March 2004. Reform Support and Project Management Unit, Water Supply and Sanitation Department, Government of Maharashtra, Maharashtra.

Sangameswaran, P., 2010. Rural drinking water reforms in Maharashtra: the role of neoliberalism. Economic and Political Weekly 45 (4), 62–69.

World Bank, 2008. Review of Effectiveness of Rural Water Supply Schemes in India. The World Bank, New Delhi.

World Bank, 2011. Jalswarajya Project: Easing the Burden of Water for Villagers in Rural Maharashtra. The World Bank, New Delhi.

World Bank, 2012. Draft Report on Maharashtra Sector Status Report: Water Supply. Water and Sanitation Program, Government of Maharashtra. The World Bank.

CHAPTER 6

Influence of Climate Variability on Performance of Local Water Bodies: Analysis of Performance of Tanks in Tamil Nadu

D.S. Kumar
Tamil Nadu Agricultural University, Coimbatore, India.

6.1 INTRODUCTION

The debate on climate change has evolved in recent years from being about whether climate change is a serious problem, toward being about when and how to address it. It is argued that climate change increases risk, particularly for those who rely on weather patterns, agriculture, water, and other natural resources for their livelihoods (El-Ashry, 2009).

Several recent studies around the globe have shown that climatic change is likely to impact significantly on freshwater availability. Changes in the global climate over the next 100 years are almost certain (IPCC, 2001), irrespective of whether they are human-induced or otherwise. Such changes will impact on the spatial and temporal distributions of surface water resources, and may well be characterized by a more frequent occurrence of extreme events (Meigh et al., 1998; Arnell and King, 1998). This clearly suggests that the availability of freshwater resources will need to be more carefully managed in future for the achievement of this goal (Sullivan et al., 2005).

In India, studies have shown that while average temperatures have increased over the last century, there is no significant trend in rainfall. It is estimated that a temperature increase of 2–3.5°C in India could result in a decline in farm revenues of between 9% and 25% (Working Group on Climate Change and Development Reports, 2004). However, rainfall has exhibited an increasing or decreasing trend at the regional level (Kavi Kumar, 2007). Mall et al. (2006) have made a review of climate change and its impacts on water resources in India. It is clear from this review that there is an increasing trend in surface temperature, and no significant trend in rainfall

Rural Water Systems for Multiple Uses and Livelihood Security
ISBN 978-0-12-804132-1
http://dx.doi.org/10.1016/B978-0-12-804132-1.00006-8

at the national level, but there are decreasing/increasing trends in rainfall at regional levels, as is evident from the several studies reviewed. In a number of studies, it is projected that increasing temperature, if accompanied by a decline in rainfall, may reduce net recharge and affect groundwater levels.

It is clear from the past studies that though there are several studies which have attempted to assess the impact of climate variability, the links between climate variability and water resources need to be thoroughly examined in order to evolve better water management strategies for agricultural production in the light of climate variability. It is in this context that the study of climate variability and its impacts on tank irrigation management assumes importance.

6.1.1 Why This Study?

There are as many obstacles to tank irrigation as there are benefits, due to their large number and the differences in water demand, managerial experiences, and investment needs for maintenance. During low-rainfall years, the tanks would store small quantities of water, and the chain of tanks, except the first tank, would receive little supply. Using 40 years rainfall data, it was estimated that in 5 out of 10 years the tanks will be experiencing deficient supply; in 3 years the tanks will fail; in 1 year the tanks will have surplus storage, and in 1 year the tanks will be getting full supply. The effect of the same would be more profound in non-system tanks resulting in a reduction in irrigated area over the years. Since 90% of the tanks are non-system tanks, the effect on area reduction would be more significant. Besides rainfall variation and tank filling, other factors such as siltation, encroachment, channel obstruction, etc., have an effect on the tank-irrigated area. The data on rainfall and area irrigated by tanks over the years show that the influence of northeast monsoon rains on the tank-irrigated area in the state was greater than the southwest rainfall. The correlation coefficient between the area irrigated by tanks and the rainfall was found to be between 0.20 and 0.30, indicating that apart from rainfall there are several other factors which have contributed to the overall decline in tank performances over the last few decades (Palanisami et al., 1997).

Other factors include heavy siltation in the tank bed, encroachment of tank beds, poor functioning of the sluices and surplus weirs of the upper tanks, severe encroachment in the supply channels, deforestation, erosion in catchments, conflicts over intertank water distribution, etc. Siltation might reduce tank water storage capacity by up to 30%, although there are cases when heavy siltation has completely eliminated the storage capacity (Palanisami and Suresh Kumar, 2004). It is reported that some tanks function only in normal/excess-rainfall years and not so in poor/low-rainfall years.

Though irrigation tanks have considerable merit in improving environmental, ecological, and socioeconomic conditions of the people who are depending on it, the declining performance of tanks and associated livelihood consequences and coping strategies have not been rigorously examined, making it difficult to assign developmental and research agenda. The key requirements for achieving developmental objectives are to examine the performance of tanks and the adaptation and coping strategies adopted by the rural farm households under different production environments. Keeping these issues in view, this chapter examines the impact of climate variability on tank irrigation and tank-based agriculture in Tamil Nadu.

6.2 TANK IRRIGATION IN TAMIL NADU

Tanks have existed in India from time immemorial and have been an important source of irrigation, especially in the southern peninsula region. Many were built in the 18th and 19th centuries by kings, Zamindars, and even the British rulers. Though they are found in all parts of the country, they are concentrated in southern states such as Andhra Pradesh, Karnataka, and Tamil Nadu. Tank irrigation systems also act as an alternative to pump projects, where energy availability, energy cost, or groundwater supplies are constraints for pumping. The distribution of tanks is quite dense in some areas.

The tanks are classified into system tanks (which receive supplemental water from major streams or reservoirs in addition to the yield of their own catchments area) and non-system or rain-fed tanks, which depend on the rainfall in their own catchments area and are not connected to major streams/reservoirs. There is also another classification based on administration.[1]

There are around 41,127 tanks in Tamil Nadu state alone, of varying sizes and types. Out of these, 81% have a command area less than 40 ha and 19% have more than 40 ha of command area. Presently, a large number of tanks are turning defunct due to various maintenance issues. While tanks are usually regarded as irrigation sources, there are several characteristics which make them well suited for multiple uses. First, tanks provide dispersed water

[1] The tanks are also classified into Panchayat Union (PU), Public Works Department (PWD), and Ex–Zamin tanks based on the management activity. The PU tanks have a command area less than 40 ha and under the control of Panchayat Unions. Tanks having a command area of more than 40 ha as well as all the system tanks are maintained by the Water Resources Department. A separate division called Ex–Zamin Standardization Division (as per Ms No. 2321 PWD DT 25.11.1982) involved in standardization of Ex–Zamin tanks having an ayacut between 10 acres (4 ha) and 100 acres (40 ha). The main task of this division are: (1) strengthening the tank bunds and bringing them to standard, (2) repairs and reconstruction of sluice, (3) repairs to the weirs, and (4) improvement and construction of leading channels.

Table 6.1 Trend in net irrigated area of different sources of irrigation in Tamil Nadu (area in lakh hectares)

Sources	1970s	1980s	1990s	2000s	2010s
Canals	8.94 (31.59)	8.30 (32.62)	8.29 (28.71)	7.57 (26.43)	6.95 (23.55)
Tanks	8.49 (30.01)	5.81 (22.86)	6.22 (21.53)	5.00 (17.46)	4.94 (16.74)
Wells	10.51 (37.14)	11.14 (43.79)	14.22 (49.23)	15.95 (55.70)	17.54 (59.46)
Others	0.35 (1.25)	0.19 (0.74)	0.15 (0.53)	0.12 (0.41)	0.07 (0.25)
Total	28.30 (100.00)	25.44 (100.00)	28.89 (100.00)	28.64 (100.00)	29.49 (100.00)

Figures within parentheses indicate percentages of the total.
Different issues of Season and Crop Report of Tamil Nadu, Department of Economics and Statistics, Government of Tamil Nadu, Chennai.

storage near many of the villages. The technology itself creates bodies of standing water that can be accessed by people and livestock. Further, tanks provide a combination of land and water resources that can be used for brick making, trees, grazing, and fish production. In water-scarce regions, tanks are used for a variety of productive and domestic uses and are therefore very important for rural livelihood (Palanisami and Meinzen-Dick, 2001).

6.2.1 Changing Sources of Irrigation

Significant changes have occurred in the irrigation landscape of the state over the years. The long-term analysis of the irrigated area in the state revealed that the net irrigated area increased from 28.30 lakh hectares during the 1970s to 29.49 lakh hectares in the 2010s. Canals and tanks were the major sources of irrigation in the 1970s (canals accounted for 31.59% and tanks accounted for 30.01%) and 1980s (canals accounted for 32.62% and tanks accounted for 22.86%), contributing to nearly two-thirds of the total net irrigated area (Table 6.1; Fig. 6.1). Currently, well irrigation is predominant in the state since it accounts for 62.86% of the total irrigated area during 2012–13, followed by canals (21.55%) and tanks (15.35%) (Season and Crop Report of Tamil Nadu, 2012–13).

It is evident that the wells have become the dominant source of irrigation since the 1980s (Table 6.2). They accounted for 37.14% in the 1970s, which increased to 43.79% in the 1980s, 49.23% during the 1990s, 55.70% during the 2000s, and finally 59.46% during the 2010s. There is a significant decline in surface irrigation (canal and tanks).

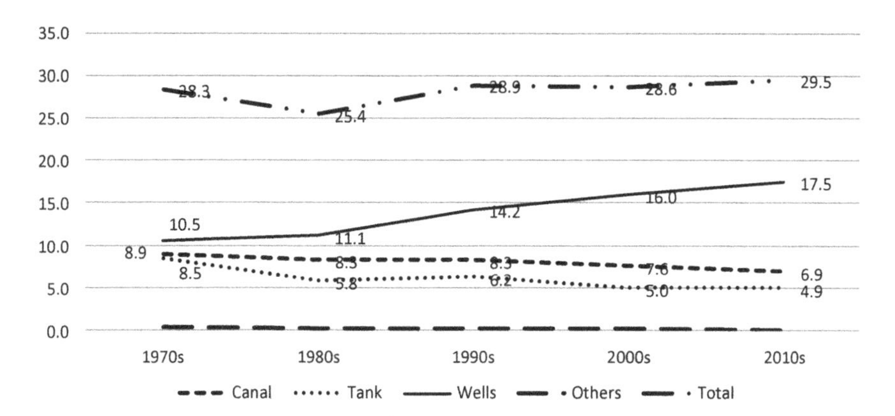

Figure 6.1 Trends in tank-irrigated area in Tamil Nadu.

Table 6.2 Growth rate in number and sources of irrigation in Tamil Nadu (percentage)

No. of sources/year	1970s	1980s	1990s	2000s	1970–2013
No. of canals	−2.13	−0.59	0.23	−0.37	−0.50
No. of tanks	0.57	−0.07	0.17	0.70	0.25
PWD (>40 ha)	−2.65	0.17	0.42	0.89	−0.0008
PU (<40 ha)	1.48	−0.12	0.12	0.65	0.33
No. of wells	0.60	1.98	0.23	−0.008	1.02
Net area irrigated	0.70	−0.22	2.27	1.26	0.22
Canals	−0.32	−0.96	0.22	0.78	−0.55
Tanks	−0.66	−1.69	1.67	0.72	−1.34
Wells	2.73	1.19	3.80	1.74	1.55
Net cropped area	−0.03	0.56	−0.49	−0.06	−0.53

Authors' own estimate based on secondary information collected from the published sources.

The growth in number of irrigation sources shows an interesting trend in the state. The number of wells has registered an annual compound growth rate of 1.02% for the period from 1970–71 to 2012–13. The growth in wells occurred at a much faster rate during the 1980s and fell slightly during the 1990s. Also, the growth shows a slightly negative trend in the 2000s. The tanks in the state have registered a positive growth rate of 0.25% over a period of five decades. The net area irrigated in the state has grown at the rate of 0.22% for the period from 1970 to 2013. Except in the 1980s, the area irrigated registered a positive growth rate.

Though the net irrigated area has registered a positive growth rate in the state, analysis of source-wise irrigated area shows a little different picture. The area under canal and tank irrigation shows a negative growth rate of 0.55% and 1.34%, respectively. The area under tank irrigation registered a negative growth rate during the 1970s and 1980s. However, consistent efforts taken by the state in investment on tank modernization under different programs and grants helped to revive tank irrigation in the state. Programs like EEC, World Bank–funded projects, TN-IAMWARM, contributed significantly to the modernization and revival of tank irrigation in the state.

Tamil Nadu is one of the tank-intensive southern states of India. The state had 41,127 tanks during the TE 2012–13, of which privately owned tanks accounted for 81% and the PWD tanks accounted for 19% (Table 6.3).

The area under tank irrigation is around 18% of the total net irrigated area in the state. Pudukottai district has the highest number of tanks (5451 tanks account for 13.3% of the total) followed by Sivagangai district (4966 tanks) and Dindigul district (3104 tanks). The districts of Ramanathapuram, Sivagangai, and Pudukottai have maximum irrigated area under tanks with reference to the total net irrigated area.

6.3 CLIMATE VARIABILITY AND TANK PERFORMANCE

The following section examines the impact of climate variability on tank irrigation management. In order to study the impacts, it is envisioned to first examine whether the variability of climatic variables like rainfall has any bearing on tank irrigation management.

6.3.1 Trend in Rainfall

Rainfall is an important factor that influences tank irrigation and in turn agriculture. The agricultural production and productivity of crops depend on the timely onset of monsoon and the quantity and distribution of rainfall. The long-term trend in rainfall in the state indicates that the state received an average annual rainfall of 927.7 mm (Table 6.4).

The northeast monsoon contributes around 50.4%, while the southwest monsoon accounts for 33.7% of the total rainfall. During the southwest monsoon, the state received an average rainfall of 312.4 mm. Average precipitation during the northeast monsoon is 467.1 mm.

The coefficient of variations demonstrates significant interannual and seasonal variations. It is observed to be around 20% for the southwest monsoon, 30% during the northeast monsoon, 134% during winter, 41.7% in summer, and 18.5% for the annual rainfall.

As the variations in rainfall are significant across regions, it may have significant implications on irrigation and cropping patterns in the state. It is evidenced that the coefficient of variations in annual rainfall has increased over the decades.

Table 6.3 Distribution of tanks across districts of Tamil Nadu during TE 2012–13 (hectares)

District	No. of tanks			% to total number of tanks in the state			Net area irrigated by tanks	Total net irrigated area	% of NIA by tanks to total NIA
	PU tanks	PW tanks	Total tanks	PU tanks	PW tanks	Total tanks			
Kancheepuram	1233	709	1942	3.7	8.9	4.7	59,985	101,685	59.0
Thiruvallur	1322	573	1895	4.0	7.2	4.6	16,866	87,337	19.3
Cuddalore	404	188	592	1.2	2.4	1.4	4700	144,979	3.2
Villupuram	1097	988	2085	3.3	12.4	5.1	55,569	233,159	23.8
Vellore	935	420	1355	2.8	5.3	3.3	1687	88,295	1.9
Thiruvannamalai	1361	605	1966	4.1	7.6	4.8	28,674	146,295	19.6
Salem	457	89	546	1.4	1.1	1.3	37	111,410	0.0
Namakkal	192	67	259	0.6	0.8	0.6	66	72,002	0.1
Dharmapuri	926	89	1015	2.8	1.1	2.5	1889	67,879	2.8
Krishnagiri	1188	139	1327	3.6	1.7	3.2	7927	51,230	15.5
Coimbatore	18	30	48	0.1	0.4	0.1	0	113,910	0.0
Erode	681	17	698	2.1	0.2	1.7	60	130,807	0.0
Tiruchirapalli	1652	115	1767	5.0	1.4	4.3	3523	91,913	3.8
Karur	248	18	266	0.7	0.2	0.6	51	54,071	0.1
Perambalur	201	51	252	0.6	0.6	0.6	2804	30,529	9.2
Pudukottai	**4791**	**660**	**5451**	**14.5**	**8.3**	**13.3**	**71,547**	**110,149**	***65.0***
Thanjavur	298	130	428	0.9	1.6	1.0	91	188,405	0.0
Madurai	1995	294	2289	6.0	3.7	5.6	19,531	74,575	26.2
Theni	130	20	150	0.4	0.3	0.4	1367	64,333	2.1
Dindigul	**2338**	**766**	**3104**	**7.1**	**9.6**	**7.5**	**8022**	**114,936**	**7.0**
Ramanathapuram	1217	477	1694	3.7	6.0	4.1	53,820	65,614	***82.0***
Virudhunagar	707	290	997	2.1	3.6	2.4	22,555	51,824	43.5

Continued

Table 6.3 Distribution of tanks across districts of Tamil Nadu during TE 2012–13 (hectares) —cont'd

District	No. of tanks			% to total number of tanks in the state			Net area irrigated by tanks	Total net irrigated area	% of NIA by tanks to total NIA
	PU tanks	PW tanks	Total tanks	PU tanks	PW tanks	Total tanks			
Sivagangai	**4288**	**678**	**4966**	**12.9**	**8.5**	**12.1**	**64,372**	**86,279**	***74.6***
Tirunelveli	1782	373	2155	5.4	4.7	5.2	41,628	106,910	38.9
Thoothukudi	544	107	651	1.6	1.3	1.6	6488	37,096	17.5
Kanyakumari	2582	41	2623	7.8	0.5	6.4	16,344	28,873	56.6
Thirupur	24	18	42	0.1	0.2	0.1	1229	112,620	1.1
Ariyalur	531	33	564	1.6	0.4	1.4	7970	32,432	24.6
Total	33,142	7985	41,127	100.0	100.0	100.0	592,780	3,148,057	18.8

NIA, net irrigated area; *PU tanks*, tanks managed by Panchayat Union which have a command area of less than 40 ha; *PWD tanks*, tanks managed by Water Resources Organization which have a command area of more than 40 ha.

Different issues of Season and Crop Report of Tamil Nadu, Department of Economics and Statistics, Government of Tamil Nadu, Chennai.

Table 6.4 Decadal changes in rainfall in Tamil Nadu

Year	1970s	1980s	1990s	2000s	2010s	Overall
South-West						
Mean	329.3	309.9	310.3	300.8	310.0	312.4
Standard deviation	40.7	78.5	71.7	53.5	69.3	61.0
Coefficient of variation (%)	12.4	25.3	23.1	17.8	22.4	19.5
North-East						
Mean	480.6	369.4	519.7	487.4	505.5	467.1
Standard deviation	149.0	92.3	153.8	137.5	121.3	140.0
Coefficient of variation (%)	31.0	25.0	29.6	28.2	24.0	30.0
Winter						
Mean	17.6	38.6	27.5	21.4	26.8	26.3
Standard deviation	18.5	59.6	34.2	20.4	15.0	35.3
Coefficient of variation (%)	104.8	154.5	124.3	95.3	56.0	134.1
Summer						
Mean	115.5	117.0	97.9	161.8	106.2	121.8
Standard deviation	34.2	36.4	43.5	70.3	29.4	50.9
Coefficient of variation (%)	29.6	31.1	44.5	43.4	27.7	41.7
Annual						
Mean	943.0	834.8	955.4	971.4	948.4	927.7
Standard deviation	130.5	184.7	172.8	183.7	211.2	171.7
Coefficient of variation (%)	13.8	22.1	18.1	18.9	22.3	18.5

Author's own estimate.
Data collected from different issues of Season and Crop Report of Tamil Nadu, Department of Economics and Statistics, Government of Tamil Nadu, Chennai.

6.3.2 Whether Rainfall Has any Bearing on Tank Irrigation?

There has been a consistent decline in tank-irrigated area since 1971–72 though there are variations in the rainfall over the years (Fig. 6.2).

The correlation between tank net irrigated area and annual rainfall in the state is 0.32 (Fig. 6.3). Though the rainfall is the major factor determining the tank irrigated area (Fig. 6.4), the correlation is low, implying that the tank-irrigated area is also determined by other factors.

In order to capture the effect of rainfall, detrended data were used. *Detrending* is the statistical or mathematical operation of removing trend from the series. Detrending is often applied to remove a feature, believed to be distorting or obscuring a relationship of interest. In climatology, for

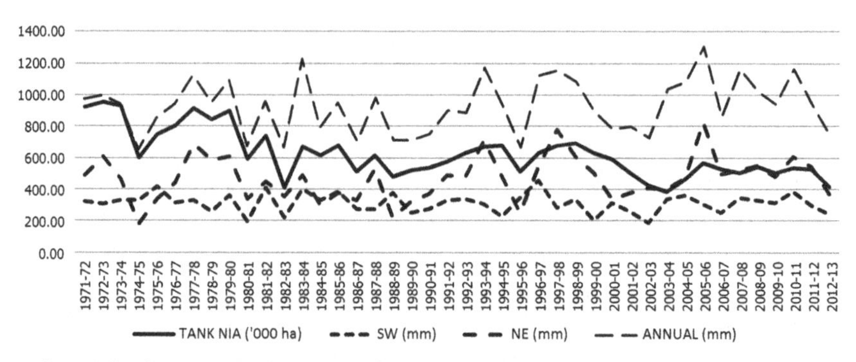

Correlation between: Tank area and Southwest: 0.24; Tank area and North-east: 0.35
Tank area and Annual Rainfall: 0.32

Figure 6.2 Trend in rainfall and area under tank irrigation in Tamil Nadu.

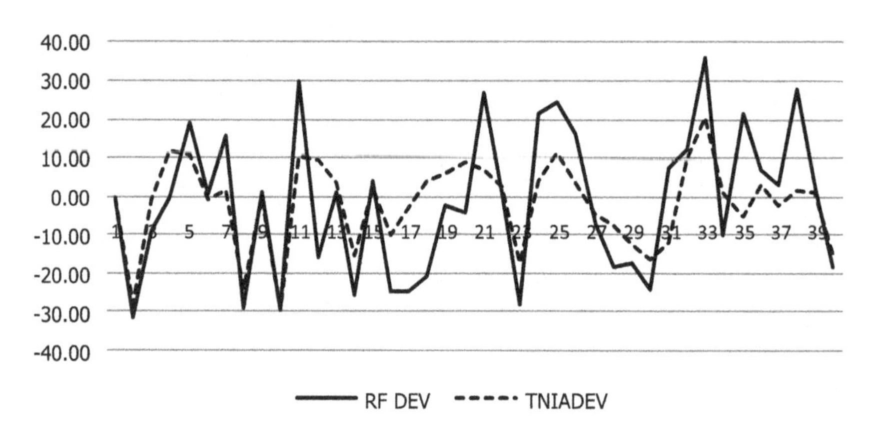

Figure 6.3 Trend in deviations in annual rainfall and net irrigated area by tanks.

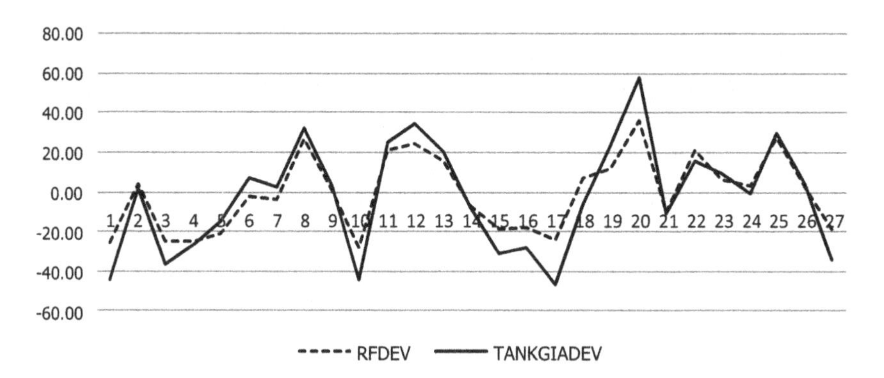

Figure 6.4 Trend in deviations in rainfall and gross irrigated area by tanks.

example, a temperature trend due to urban warming might obscure a relationship between cloudiness and air temperature. Detrending is also sometimes used as a preprocessing step to prepare time series for analysis by methods that assume stationarity. This type of removing a trend from the data enables us to focus our analysis on the fluctuations in the data about the trend. A linear trend typically indicates a systematic increase or decrease in the data. A systematic shift can result from sensor drift, for example. While trends can be meaningful, some types of analyses yield better insight once we remove trends. Whether it makes sense to remove trend effects in the data often depends on the objectives of our analysis. *Trend* in a time series is a slow, gradual change in some property of the series over the whole interval under investigation. Trend is sometimes loosely defined as a long-term change in the mean.

For the purpose of analysis, it is assumed that the tank-irrigated area is not only influenced by the rainfall but also the management factors (such as poor maintenance of tanks, silted tanks, etc.) and human-induced factors (such as encroachment of tank beds, supply channels, blocking of supply channels, etc.). Hence, in order to study the relationship between the effect of rainfall on tank-irrigated area, detrending of tank-irrigated area was employed. Secondly, we have also tried to map the relationship between rainfall and tank-irrigated area using the deviations in rainfall and tank-irrigated area (rainfall anomalies). The trend in deviations in annual rainfall and tank net irrigated area follows a certain definite pattern. The correlation is worked out to be 0.73. The correlation between the deviations in annual rainfall and tank gross irrigated area is 0.67 and the pattern of variations follows a uniform deviation.

It is clear from the analysis that the area irrigated by tanks is not only determined by the rainfall but also by other factors such as encroachment, urbanization, growth in the number of wells, demand for land for non-agricultural uses, etc. To examine whether these factors really matter, we tried to map the relationship between growth in tank irrigation in the state and other human-induced factors (Fig. 6.5).

The area under tank irrigation has been continuously declining over the years. In contrast, the area under well irrigation has been constantly increasing over the same time period. The first shortfall in the area irrigated by tanks was observed during 1974–75, when the area declined to 5.93 lakh hectares from 9.3 lakh hectares in the previous year. This was mainly attributed to monsoon failure. The total annual rainfall was 647.4 mm. The other drivers, namely the population density and land put

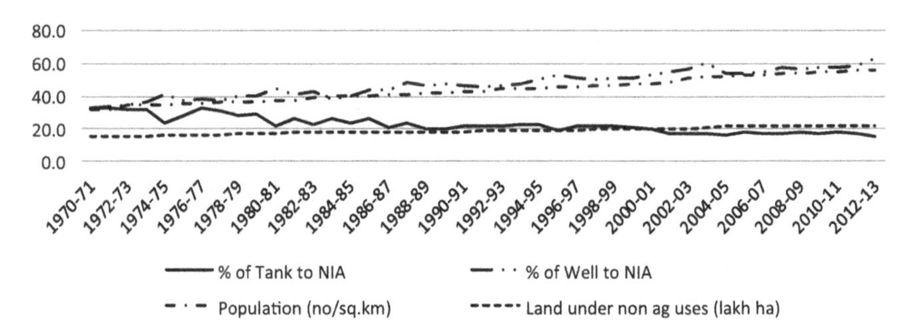

Figure 6.5 Trend in tank, well irrigation, land under non-agricultural uses and population density.

under non-agricultural uses, have been continuously increasing over the years. Thus, there is a need to study the factors which determine the tank irrigation in the state.

6.3.3 Factors Determining Tank Irrigation at State Level

To understand whether rainfall has any relationship on tank irrigation, a multiple regression was set up using tank performance index[2] as the dependent variable. The following model was made:

$$\begin{aligned}\text{TANKPER} &= a_0 + a_1\text{NWELLS} + a_2\text{RAINFALL} + a_3\text{NAGLAND}\\ &\quad + a_4\text{POPDENSITY} + a_5\text{TREND} + U_i \qquad [6.1]\end{aligned}$$

The tank performance, which is the dependent variable, is expected to be influenced by number of well (NWELLS), rainfall in mm (RAINFALL), land under non-agricultural uses in ha (NAGLAND), population density in number/sq. km (POPDENSITY), and TREND. The dependence on tank water is an important factor that affects the tank performance. The number of wells is included to capture the effects of resource dependence. The greater the number of wells in the command area of the tank, the less would be the dependence of farmers on tank irrigation. An adverse consequence of this would be that there would be no incentive for farmers to contribute to labor and other costs for tank management and maintenance. The storage capacity of the tanks would decline, affecting the tank performance. Thus, it

[2] Tank performance is generally measured as the ratio of actual area irrigated by the tank to the total command area.

$$\text{Tank Performance} = \frac{\text{Actual Area Irrigated by Tank}}{\text{Total Command Area of Tank}}$$

is expected that the number of wells is expected to influence negatively the tank performance.

Rainfall directly affects tank water storage and irrigation potential. Rainfall is expected to positively influence tank performance. Another important factor which influences tank performance is urbanization and the demand for land for non-agricultural uses. In the process of urbanization, conversion of land for non-agricultural purposes takes place at a faster rate, thus reducing tank water spread, catchment area, and area under tank irrigation. Thus, non-agricultural use of land due to urbanization can have a negative influence on tank performance. Population density is included in the model to capture the effect of encroachment to catchment area, tank bed and tank water spread area, supply channels, etc., all of which can reduce the tank performance.

The TREND is included mainly to examine the effects of management and maintenance activities undertaken continuously over the years. Over the years the state government with the help of the central government and international donors has made huge investments on tank maintenance activities. In addition, the Water Users' Associations (WUAs) look after maintenance activities. These management and maintenance activities help improve the conditions of the tanks.

This ensures good storage of water and is expected to result in improved tank performance. Thus, the variable TREND is expected to influence the tank performance positively. The estimated results indicate that the tank performance is significantly influenced by rainfall (RAINFALL), land under non-agricultural uses (NAGLAND), population density (POPDENSITY), and TREND as expected (Table 6.5).

Table 6.5 Factors influencing tank performance

Variables	Coefficients	Std Error	"*t*" ratio
CONSTANT	400.86	65.034	6.16
NWELLS	−0.0000006	0.000009	−0.06
RAINFALL	0.056743***	0.0068	8.31
NAGLAND	−0.00013***	0.00004	−3.77
POPDENSITY	−0.4712***	0.1645	−2.86
TREND	3.7415***	0.9973	3.75
Adjusted R-Squared	0.82		
F-statistics	38.71***		

***, $P < 0.01$ indicates significant at 1% level.

6.4 IMPACT OF CLIMATE VARIABILITY ON TANK-BASED AGRICULTURE

6.4.1 Profile of Study Tanks

To study the impact of rainfall on tank irrigation management, different domains of impacts were studied. For the purpose two tanks in Villupuram district were studied (Table 6.6). The Alampoondi and Thiruvampattu tanks of Villupuram district also represent the well-dominated situation. Both tanks are non-system tanks and the command area is 209 and 274 ha, respectively. The major crop is rice, followed by sugarcane and groundnut. The well density is 0.24 and 0.38, respectively (Fig. 6.6).

Table 6.6 Profile of the study tanks in Villupuram district

Particulars	Alampoondi	Thiruvampattu
Registered command area (ha)	209	274
System/non-system	Non-system	Non-system
Number of wells in the command area	50	105
Well density (no of wells/ha)	0.24	0.38
Number of farmers		
Marginal (<1.0 ha)	350 (43.8)	220 (44.4)
Small (1–2 ha)	300 (37.5)	150 (30.3)
Medium (2–4 ha)	100 (12.5)	75 (15.1)
Large (>4 ha)	50 (6.3)	50 (10.1)
Total	800 (100.0)	495 (100.0)
Average size of holding (ha)	0.86	0.68
Major crops	Rice, sugarcane	Rice, sugarcane

Water Resources Department and Village Administrative Offices of the concerned tanks and villages.

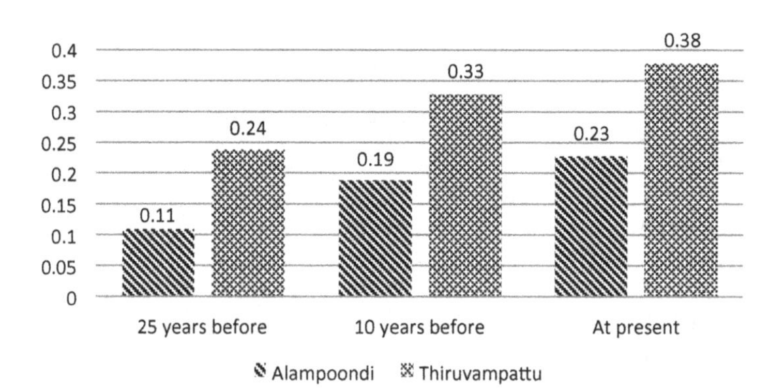

Figure 6.6 Trend in well density.

6.4.2 Tank Level Impacts

6.4.2.1 Tank Performance

To understand the impact of variability at the tank level, the area actually irrigated by tanks and the tank performance were examined. Table 6.7 gives the basic statistics of the study tanks across districts. The analysis of 34 years of data from 1980 to 2013 for these tanks reveals that the mean tank performance is 70% and 76% respectively, for Alampoondi and Thiruvampattu tanks.

6.4.2.2 Impact on Cropping Pattern

The tanks were constructed in ancient days mainly to cultivate rice (either single or double crop). The purpose was to harvest and store rainwater in the tanks and use it for rice cultivation. In addition to cultivation of crops, tank water has been used for rearing fish, trees, domestic purposes, and other environmental services (Table 6.8).

The variability in climate is expected to reduce water availability in tanks and in turn the different services provided by the tanks. Reduced water availability in the tanks will result in reduced cropped area, reduction in other activities such as fishing, lotus collection, and other provisioning services.

It is observed that the major impact of such variability is reduced or there is no cropped area under rice. This negative impact is mainly due to reduced water availability as a result of climate variability. Twenty years ago, crops were grown in two seasons, but now the crop is grown only in one season. In addition to rice, other crops such as sugarcane, groundnut, and other pulses are also grown.

Table 6.7 Details of tank-irrigated area and tank performance

Particulars	Unit	Alampoondi tank	Thiruvampattu tank
Command area	ha	209	274
Actual area irrigated by tank			
Mean	ha	146	209.4
Std deviation		57.0	76.2
CV	%	39.0	36.4
Tank performance	%		
Mean	ha	70.0	76.4
Std deviation		27.0	27.8
CV	%	39.0	36.4

Author's own estimates.
Water Resources Department, Villupuram.

Table 6.8 Impact on cropping in Alampoondi and Thiruvampattu tanks (area in hectares)

	Alampoondi tank			Thiruvampattu tank		
Crops	**25 years before**	**10 years before**	**At present**	**25 years before**	**10 years before**	**At present**
	220.0 (76.7)	172.0 (66.2)	129.8 (56.4)	350.0 (74.5)	300.0 (69.0)	150.0 (41.0)
Sugarcane	20.0 (7.0)	34.0 (13.1)	44.5 (19.3)	35.0 (7.4)	35.0 (8.0)	82.0 (22.4)
Groundnut	20.0 (7.0)	29.0 (11.2)	28.0 (12.2)	–	–	–
Others	27.0 (9.4)	25.0 (9.6)	28.0 (12.2)	85.0 (18.1)	100.0 (23.0)	134.0 (36.6)
Total	287.0 (100.0)	260.0 (100.0)	230.3 (100.0)	470.0 (100.0)	435.0 (100.0)	366.0 (100.0)

Figures in parentheses indicate percentage of the total.

To cope with climate variability, farmers in the command area altered their cropping pattern. There is a reduction in area under rice (a water-intensive crop) with an associated increase in the area under sugarcane. Across all tanks a significant reduction in area under rice is noted with a consequential increase in the area under high-valued crops. Farmers in regions where there are wells in tank command areas have shown an affinity toward crops like sugarcane, vegetables, etc. However, in the deficit seasons, crops like groundnut, pulses (black gram), and cotton are grown under unirrigated conditions.

In addition, changing of rice varieties from long-duration to medium- and short-duration varieties, delayed sowing and reduced number of irrigations are being adopted by farmers. Adoption of microirrigation, particularly drip irrigation, allowing tanks for groundwater recharge (at times of scarcity), and livelihood diversification (livestock, off-farm and non-farm activities) are other coping strategies followed by farmers.

6.4.2.3 Impact on Groundwater Dependence

Over the years growing water scarcity, coupled with monsoon failure, poor water availability in tanks, and insufficient water for irrigation has forced farmers in tank command areas to drill new wells and bore wells in order to cope with reduced water availability. This is one of the important adaptation strategies being followed by the farmers in tank command areas. Construction and drilling of wells is found to be common practice across these regions. It is worthwhile to note that the number of wells in tank command areas has increased over the years.

The well density has increased in both tank command areas. The well density has increased from 0.11 and 0.24 to 0.23 and 0.38 in Alampoondi and Thiruvampattu tanks, respectively. This clearly shows that climate variability led to an increase in well investment in the tank command areas.

6.4.3 Household-Level Analysis

For the purpose of examining the impact of climate variability at the household level, farm households and the agricultural landless labor households were selected employing a simple random sampling procedure. In a tank, a total of 120 farm households and 40 landless agricultural labor households were selected for the purposes of the study.

The general characteristics of farm households across study tanks are presented in Table 6.9. The average size of a farmholding is less than 1 ha in both the tanks. The cropping intensity, which indicates the intensity of the land put under use, is around 100%.

The net irrigated area, which is the maximum area which could be irrigated in particular seasons, is around half a hectare in both tanks. Both the cropping and irrigation intensity are higher (more than 100%) in tanks where there is supplementary irrigation by wells. Farmers with access to irrigation by wells (bore well or open well) are able to grow multiple crops in a year.

6.4.3.1 Well Investment

Non-availability of tank water coupled with growing groundwater scarcity and cheap power supply resulted in further degradation of the groundwater resource in the state (Table 6.10). It is argued that inefficient pricing of electricity has reduced the marginal costs of water, as well as electricity, to zero. As a result, farmers use both groundwater and electricity inefficiently. The effect of such cheaper electricity has resulted in negative externalities such as overpumping, changes in crop pattern toward more water-intensive crops, well deepening, increase in well investments, pumping costs, well failure, and abandonment and outmigration which are all increasing at a fast pace.

To cope with the non-availability of tank water supply, farmers make huge investments in groundwater extraction. These include investment on drilling new bore wells or dug wells, deepening of existing wells,

Table 6.9 General characteristics of farm households in the study area

Name of the tank	Farm size (ha)	Net sown area (ha)	Gross cropped area (ha)	Cropping intensity[a] (%)	Net irrigated area (ha)	Gross irrigated area (ha)	Irrigation intensity[b] (%)
Alampoondi	0.78	0.67	0.70	104.5	0.50	0.59	118.0
Thiruvampattu	0.63	0.58	0.60	103.4	0.40	0.42	105.0

[a]Cropping intensity is defined as the ratio of gross cropped area to net sown area and expressed as a percentage.
[b]Irrigation intensity is the ratio of gross irrigated area to net irrigated area and expressed as a percentage.
Farm household survey, 2013–14.

Table 6.10 Investment on groundwater abstraction structures (rupees/ha)

Tank name	Investment on wells	Investment on electric motor	Total amortized cost
Alampoondi	23,932.2 (76.5)	7342.8 (23.5)	31,275.0 (100.0)
Thiruvampattu	26,476.9 (77.0)	7898.6 (23.0)	34,375.5 (100.0)

Farm household survey, 2013–14.

construction of intermediate storage structures and microirrigation technologies like drip irrigation, sprinkler irrigation, and so on. Thus, the investment on irrigation structures is crucial to the study. The total amortized cost of irrigation investment is worked out as the sum of amortized cost on wells, electric motor and equipment, surface storage tanks, and drip irrigation equipment.[3]

The analysis on well and irrigation investments revealed that the total amortized cost of irrigation structures is more than Rs 30,000/ha. The reason for this huge cost is mainly the increased costs of drilling new bore wells and well-digging costs, non-availability of laborers for well digging, etc. Of the total fixed investments, the investment on wells assumes three-fourth of the total cost.

6.4.3.2 Irrigation Details

Being the ayacutdars, farmers rely mainly on tank water for irrigating the dominant crop. At times of water scarcity, farmers depend on well water for crop production. Thus, irrigation through wells forms a good supplement to tank water, especially at times of water scarcity during critical stages of crop growth. Farmers in the command area generally use well water for growing other crops.

Farmers irrigate their crops based on availability of water in the tank, as the tank water supply has reduced in recent years.

[3] The amortization of irrigation structures as follows:

$$\textit{Amortized cost of well} = \left[(\textit{Compounded cost of well}) * (1+i)^{AL} * i\right] / \left[(1+i)^{AL} - 1\right]$$

Where

AL = Average life of wells

Compounded cost of well = (Initial investment on well)*$(1 + i)^{(2014\text{-year of construction})}$

The discount rate of 5% is used in amortization reflecting long-term sustainable rate. Similarly investment on conveyance, pump set, and electrical installation were amortized. Where AL is average life of wells and it is assumed to be 30 years based on the average life of well life in the study area. Similarly, the average life of electrical motors is assumed to be 15 years.

On average, the well owners provide three to four supplemental irrigations to their crops from the wells. Water scarcity forces the farmers to buy water from the well owners in order to prevent huge losses. The practice of buying water is commonly observed in the majority of the tanks. The tank water supplemented by well water is crucial for growing rice in the region.

Farmers in the command area use tank water for irrigation. On average farmers in the tank command area apply around 10,000 m^3/ha for rice crop (Table 6.11). It is interesting to note that farmers in general use less than the crop water requirement of 13,000 m^3/ha. This is due to supplementation by rainfall and inadequate tank water supply.

The yield of crop in a region is an indicator of the performance of the system. The yield per hectare is worked out to be around 5 tons/ha in both tanks. The well owners get an increase in yield of 14% and 16%, respectively, in the Alampoondi and Thiruvampattu tanks.

The average productivity of water is crucial. The yield per m^3 of water is worked out to around 0.47 kg/m^3 of irrigation water applied. The yield/m^3 of water is found to be little higher among well owners when compared to non-well owners. This indicates that when compared to non-well owners, the well owners operate little more efficiently. This is mainly because of the supplemental irrigation provided by the wells at times of water scarcity and inadequate tank water supply. Thus, we can conclude that supplemental irrigation provided by wells enables the farmers to get increased yield of rice and operate more efficiently when compared to non-well owners in the light of climate variability.

6.4.3.3 Value of Marginal Product of Water

The economic value of irrigation water was determined by employing a production function approach (Gibbons, 1987). The marginal value of water is the marginal physical product times the output price. A quadratic production function was estimated with yield (kg/ha) as a dependent variable and volume of irrigation water used in ha·cm (WATER) as independent variable. The estimated production function is as follows:

$$\text{YIELD} = a_0 + b_1\text{WATER} + b_2\text{WATER}^2 \qquad [6.2]$$

The marginal value product of water is evaluated at mean values of water use. The VMP is worked out to Rs 431.0/ha cm and 448.54/ha cm of water in Alampoondi and Thiruvampattu tanks, respectively (Table 6.12).

Table 6.11 Irrigation water use and yield of rice in sample farms

Name of the tank	Water used (m^3/ha)			Yield (kg/ha)			Yield (kg/m^3 of water)		
	Well owners	Non-well owners	All farms	Well owners	Non-well owners	All farms	Well owners	Non-well owners	All farms
Alampoondi	10,913.7	9596.6	10,436.3	5289.1	4602.6	5040.3	0.48	0.47	0.47
Thiruvampattu	10,936.6	9512.8	10,242.5	5384.6	4645.0	4972.8	0.49	0.48	0.48

Farm household survey, 2013–14.

Table 6.12 Estimated results of water production function and value marginal product of water

Name of the tank	Estimates of water production function				MPP_w	Price of paddy	VMP_w
	Constant	WATER	WATERSQ	Adj. R^2			
Alampoondi	1443.894 (2.105)	45.233** (2.547)	−0.051 (−0.309)	0.66	34.42	12.53	431.04
Thiruvampattu	1170.129 (6.504)	47.871** (2.263)	−0.058 (−0.0647)	0.70	35.81	12.52	448.54

Figures in parentheses indicate estimated "*t*" values.**, significant at 5% level; *MPPw*, marginal product of water; *VMPw*, value marginal product of water.

6.5 COPING AND ADAPTATION STRATEGIES

6.5.1 Farmers' Perceptions About Climate Variability and Its Impacts on Tank Water Supply

Water shortages for agriculture were perceived as a critical issue in all the tanks within the scope of the study (Table 6.13). It is clear from the survey that when respondents were asked to rank responses on climate variability from "strongly agree" to "strongly disagree" more than 90% chose "strongly agree" for the reasons identified. This confirms that farmers accept there is a variation in climatic factors and consequent negative impacts on tank-based agriculture.

To cope with the climate variability, farmers in different typologies adopt various measures. This section involves identification of different adaptive measures and coping strategies followed at different levels in the past and present at household and community levels. This helps to identify adaptive measures and coping strategies for different regions.

6.5.2 Growth in Number of Wells

In general, to face the growing issue of water scarcity, farmers tend to deepen their existing wells. Well deepening is an expensive initiative and is associated with a greater risk of water availability. Hence, most farmers, including marginal and small farmers install bore wells. The same is noted in the study area. However, large farmers tend to cope better with the issue of groundwater scarcity by installing a higher proportion of bore wells.

It is evident that well irrigation is an attractive form of supplemental irrigation for farmers in the tank command area. Though there is a significant increase in wells in the tank commands over a period of time, it is worth mentioning that most of the open wells were constructed before the 1980s. The drilling of bore wells attracted the tank command farmers for the following reasons: (1) the digging of new open wells or deepening of

Table 6.13 Farmers' perception about climate variability and its impacts on tank irrigation

Particulars	% of farmers
Reduction in tank water availability	100.0
Tanks fail frequently	94.0
Reduction in cropped area (rice)	92.0
Crops experience water stress	94.0
Crop failure occurs due to water stress	92.0
Lack of water for agriculture is the major problem	94.0

Table 6.14 Growth in number of wells in the study area

	Before 1980s		1980s		1990s		2000s	
Name of the tank	**Open well**	**Bore well**	**Open well**	**Bore well**	**Open well**	**Bore well**	**Open well**	**Bore well**
Alampoondi	25	–	25	7	1	7	–	9
Thiruvampattu	39	–	2	4	–	11	–	14

Farm household survey during 2013.

existing open wells was an expensive affair, (2) non-availability of laborers for well digging, and (3) it is believed that bore wells enable the farmers to draw water from deeper levels (Table 6.14).

6.5.3 Water Market

Though there is no formal water market, farmers in the tank command area in the study region purchase water from well owners at times of water scarcity. This is a common phenomenon in all the study tanks. The water market is slightly more active in regions where the number of wells is higher. Even though there are instances of rich farmers buying water, but generally economically weaker farmers purchase water.

There are a set of factors which are responsible for purchase of water for agricultural crop production activities (Table 6.15). Inadequate water supply in the tank forms the major factor followed by no well, lack of capital for investment in wells, and saving the crop from drought form the key factors influencing water purchases.

6.5.4 Other Coping Strategies Followed

In addition to construction of wells/bore wells and construction of farm ponds, farmers adopt various other coping strategies. Practices like changing

Table 6.15 Factors responsible for water purchase at farm level

Particulars	Mean score	Rank
Inadequate water supply from tank	91.33	I
No well	84.26	II
Investment on wells is expensive	71.88	III
At least to save crop	62.71	IV
Traditionally followed	51.67	V
Location of the farm	35.56	VI
Inadequate water in the wells	26.93	VII

Figures are worked out mean score from Garett ranking technique.
Farm household survey during 2013–14.

Table 6.16 Crop-based coping and adaptation strategies followed by farmers across tank typologies (%)

Particulars	Alampoondi tank	Thiruvampattu tank
Varietal change in rice	80.0	81.0
Altering planting date of rice	65.5	63.0
Reduction in number of irrigation	17.5	12.5
Adoption of microirrigation	12.4	17.5
Altering cropping pattern	20.9	26.7
Crop diversification	21.0	26.0
Fallow lands (no cultivation)	27.5	22.5

Farm household survey during 2013–14.

from long-duration to medium- and short-duration varieties, delayed sowing, reducing number of irrigations are followed for rice. Shifting toward perennial crops like coconut, allowing tanks for groundwater recharge (at times of scarcity), livelihood diversification (off-farm and non-farm activities) are other strategies which are looked at. The details of other coping strategies followed by farmers are discussed below (Table 6.16).

The climate variability leads to a reduction in tank water availability. This has motivated the farmers in the tank command area to adopt various coping and adaptation strategies. Being a wetland ecosystem, one of the major impacts is a reduction in the cropped area, particularly that under rice. Normally in tank command areas of Tamil Nadu, one rice crop during the months of October to January is cultivated. Inadequate water supply in the tanks compels the farmers to grow short-/medium-duration varieties like Super Ponni, Co-46, and ADT 36. In almost all the tank commands, change in the variety of rice has been seen.

Depending on the availability of water in the tank and onset of monsoons, farmers alter sowing dates of rice crop in the tank command area. Delayed sowing is a common practice followed by farmers. Inadequate water supply forces the farmers to adopt other coping strategies like reducing the number of irrigations, adoption of water management technologies like direct seeding, partial or full adoption of system of rice intensification, alternate wet and dry, etc. Adoption of microirrigation, particularly in the cultivation of sugarcane, coconut, and sometimes vegetables has also come to light recently. Adoption of microirrigation, particularly drip irrigation, is becoming popular in the tank commands where well irrigation assumes importance. Well owners who have assured water supply adopt microirrigation.

Altering cropping pattern and farm diversification, like inclusion of livestock, is another strategy. This solves not only the cash flow problems but also provides employment. Marginal and small farmers with no access for wells leave the land fallow at times of non-availability of water.

6.6 CONCLUSION AND POLICIES

The study on the impact of climate variability on tank irrigation has brought out important observations that would educate policymakers to make appropriate decisions for sustainable management of irrigation tanks in the state. The major conclusions and identified policy options are discussed below.

- Rainfall has a significant bearing on the performance of tanks in the state. In addition, factors such as population density, land put under non-agricultural uses are also found to be significantly influencing tank performance. It is seen that the area under rice cultivation has reduced significantly across the region. Thus, farmers need adequate support in terms of supply of high-yielding, medium- and short-duration, drought-tolerant rice varieties. The analysis, however, conclusively shows that continued marginalization of a majority of holdings and an increasing number of marginal and small-holdings would only make farming increasingly risky with changing climatic conditions, necessitating some innovative approaches such as cluster approach (commodity groups), promoting contract farming, credit and insurance coverage to make farming viable and farmers empowered to survive in the changing scenario.
- The study found that wells in the tank command areas are very effective for supplemental irrigation, which helps save the crops and increases the yield and returns. Hence, it is suggested that adequate support may be extended to farmers in the tank-dominated situations, particularly in the non-system tank commands for construction of wells. However, it should be carefully noted that the number of wells should not exceed the threshold level.
- **Microirrigation**: One of the important water management practices followed by farmers is microirrigation. Many studies have found that microirrigation, particularly drip irrigation, results in a significant resource saving, reduction in cost of cultivation, and enhancement of crop yields and farm profitability. Hence, continuing public support for the wider adoption of microirrigation technologies appears warranted. For speedy growth of microirrigation in the state, the financial institutions may be geared up to offer special loans for its installation. This will encourage even the resource-poor marginal and small farmers to go for adoption of microirrigation.

- Water Users' Associations (WUAs) involved in management and maintenance of tanks are yet to become more proactive. Considering the importance of various institutions, in relation to tank management, defining the roles of different organizations is crucial at the initial stage so as to achieve sustainable management. This will facilitate developing linkages between different organizations involved in natural resource management, tank management in particular, resolve conflicts, and promote proper maintenance and management of tanks. Stable and sufficient financial resources are crucial for better long-term planning and sustainable management of natural resources.
- **Research**: Research organizations must be encouraged to evolve crop varieties and water management technologies that suit different types of soils and tank typologies. Research on the effect of irrigation and sustainability on yields under various water-saving methods and irrigation technologies may be encouraged. Exploratory and in-depth socioeconomic research is highly warranted to identify the extent of awareness and knowledge about climate change impacts, adaptation, constraints in adoption of various coping and adaptation strategies, transaction costs in technology adoption, and to identify policy options for various tank typologies.
- **Capacity-building**: Though farmers are aware of the impact of climate variability, coping, and adaptation strategies, there is a lack of awareness among them about water management technologies, irrigation scheduling, best agricultural practices, etc. Thus, there is a dire need for building the capacity of the farming community. Implementation of proper educational and training programs for farmers with an emphasis on the major issues, in the involvement of users of water on mitigation of drought and floods, and other extreme events. Also, adequate technical support in water management technologies and cultivation of crops, cropping pattern, and crop allocation decisions will help them better cope with climate variability.

REFERENCES

Arnell, N.W., King, R., 1998. Implications of Climate Change for Global Water Resources. Report to Department of the Environment. Transport and the Regions by Department of Geography, University of Southampton, UK.

El-Ashry, M., 2009. Adaptation to Climate Change: Building Resilience and Reducing Vulnerability, Recommendations from the 2009 Brooking Blum Roundtable. United Nations Foundations.

Gibbons, D.C., 1987. The Economic Value of Water. Resources for the Future, Washington, DC.

IPCC, 2001. Intergovernmental panel on climate change, climate change 2001. Impacts, adaptation and vulnerability. In: McCarthy, J.J., Canziani, O.F., Leary, N.A., Dokken, D.J., White, K.S. (Eds.), Contribution of Working Group II to the Third Assessment Report of the Intergovernmental Panel on Climate Change. Cambridge University Press, Cambridge, UK. ISBN: 0-521-80768-9. 1032 pp.

Kavi Kumar, K.S., November 2007. Climate change studies in Indian agriculture. Economic and Political Weekly 17, 13–18.

Mall, R.K., Gupta, A., Singh, R., Singh, R.S., Rathore, L.S., 2006. Water resources and climate change: an Indian perspective. Current Science 90 (12), 1610–1626.

Meigh, J.R., McKenzie, A.A., Austin, B.N., Bradford, R.B., Reynard, N.S., 1998. Assessment of Global Water Resources – Phase II, Estimates of Present and Future Water Availability for Eastern and Southern Africa. DFID Report 98/4. Institute of Hydrology, Wallingford.

Palanisami, K., Meinzen-Dick, R., 2001. Tank performance and multiple uses in Tamil Nadu, South India. Irrigation and Drainage System 15, 173–195.

Palanisami, K., Suresh Kumar, D., 2004. Study About Suggestions for Tank Water Sharing in Kappiyampuliyur, Vakkur Tanks in Villupuram District and Chengam Tank in Thiruvannamalai District. Report Submitted to Institute for Water Studies, Chennai.

Palanisami, K., Balasubramanian, R., Mohamed Ali, A., 1997. Present Status and Future Strategies of Tank Irrigation in Tamil Nadu. Tamil Nadu Agricultural University, Coimbatore.

Sullivan Caroline, Jeremy Meigh, and Peter Lawrence (2005). Application of the Water Poverty Index at Different Scales: A Cautionary Tale, Agriculture Ecosystem and the Environment, 2005.

Working Group on Climate Change and Development Reports, 2004. Up in Smoke: Threats From, and Responses to, the Impact of Global Warming on Human Development.

CHAPTER 7

Groundwater Use and Decline in Tank Irrigation? Analysis From Erstwhile Andhra Pradesh

M.D. Kumar
Institute for Resource Analysis and Policy, Hyderabad, India

N. Vedantam
Engineering & Research International LLC, Abu Dhabi, United Arab Emirates

7.1 INTRODUCTION

Tanks have been an important source of irrigation in India for generations. The states of Andhra Pradesh, Karnataka, and Tamil Nadu have the largest concentration of irrigation tanks, numbering 120,000 (Palanisami et al., 2010), and accounting for nearly 60% of India's tank-irrigated area (Karthikeyan, 2010). They play the vital role of harvesting surface runoff during monsoon and then allowing it to be used later. The predominance of tanks in the Deccan plateau is because of the unique topographic characteristics of the regions. The areas falling under these regions offer ideal potential for tank construction and carrying out gravity-based irrigation (ADB, 2006).

Tanks are very important from an ecological perspective as they help conserve soil, water, and biodiversity (Balasubramanian and Selvaraj, 2003). In addition, tanks also contribute to groundwater recharge, flood control, and silt capture (Mosse, 1999). Although most of the tanks were essentially constructed for irrigation purposes, they have also been used for providing water for domestic and livestock consumption. Over the years, the multiple-use dependence on tanks has only increased. The tank-irrigated area has been declining in India over the years. In Andhra Pradesh, figures are more alarming with the net area irrigated by tanks reducing to 490,000 hectares (ha) in 2003–04 from 747,000 ha in 1995–96, a decline of 35% (Source: Ministry of Agriculture, Govt. of India).

The impact of decline in tank systems on the rural communities, who have been traditionally dependent on them, is manifold as their dependence

Rural Water Systems for Multiple Uses and Livelihood Security
ISBN 978-0-12-804132-1
http://dx.doi.org/10.1016/B978-0-12-804132-1.00007-X

is not only for water for irrigation and domestic use but also for forestry, fisheries, brick-making, manure, and fodder. The neglect of tanks has resulted in farmers receiving insufficient quantities of water from tanks (Palanisami, 2006). A study conducted in tank-irrigated areas of Tamil Nadu estimated reduction in crop yield and income for tank-dependent farmers owing to the growth of private well irrigation and a deterioration in tank performance (Kajisa et al., 2004). Thus, a well-functioning tank system has a significant bearing on the household income, especially for the small farmers who have limited private resources to invest in wells and pump-sets.

Recent attempts to modernize and rejuvenate existing irrigation tanks have focused more on physical rehabilitation, with little or no emphasis on the understanding of tank hydrology. Particularly, the way land-use changes in the catchment are affecting tank inflows and siltation rates, etc., have been paid least attention by those who are involved in tank rehabilitation programs. The fact is that intensive crop cultivation, often in the common land through encroachments, and intensive pumping of groundwater in the upper catchments for irrigation are likely to threaten the very sustainability of the tank ecology in many areas. Intensive cultivation will impound a significant share of the catchment runoff; whereas excessive groundwater pumping in hilly areas can reduce groundwater outflows into streams, which constitute part of the tank inflows downstream.

If this is so, it will have serious implications for tank management programs. From a physical systems perspective, if the performance of tanks is to be sustained or improved, it will be important to influence the land-use decisions and groundwater use in the catchment. From an institutional perspective, the domain of the conventional institutions that are being created to manage the tanks by governments and NGOs alike will have to expand to bring groundwater users and catchment cultivators under its fold. This calls for developing entirely new sets of protocols for tank rehabilitation, including physical strategies for tank management and institutional arrangements for ensuring their sustainable performance.

7.2 CAUSE OF DECLINE OF TANKS: CONTESTED TERRAINS?

A large volume of literature exists on the "decline" of tanks in south India. This "decline" is both in terms of declining relative contribution of tanks in irrigated areas, and reduction in aggregate area under tank irrigation (Sharma, 2003). However, the theory of "decline of tanks" is contested by a few scholars, who believe that this emerges from a very reductionist approach of viewing tanks as mere sources of irrigation, and that the criteria for evaluating the performance of tanks should be more broad, to accommodate their various social, economic, and ecological functions (Kumar, 2002; Palanisami et al., 2010). Kumar (2002), for instance, argues that given the wide range of physical, social, economic, and ecological functions which tanks perform, the simplistic criteria for evaluating the performance of tanks, which look at the irrigated area or the number of users of drinking water as the indicators, need to change and more complex criteria need to be evolved.

Nevertheless, various theories have been made as to the reason for the "decline of tanks" in south India. Shankari (1991) points out that poor management is primarily responsible for the decline of tanks, as evident in the non-participation of farmers in cleaning channels, encroachment of the tank bed, inadequate repairs, weed infestation, and siltation (Shankari, 1991). von Oppen and Rao (1987) argued on the basis of empirical analysis that increases in population density resulted in deforestation in catchment areas leading to soil erosion and siltation. After Sekar and Palanisami (2000), tank bed cultivation and the lack of an administrative structure to provide timely repair and maintenance contributed to the decline of tank irrigation.

The other reasons provided by scholars include: agricultural encroachment of supply channels and tank beds, which reduced the inflows into the tanks (Easter and Palanisami, 1986; Mosse, 1999); sand mining of supply channels; rural infrastructure development interfering with the natural inflows; and unplanned watershed development cutting off the supply to tanks (ADB, 2006; Palanisami, 2006); decline in tank storage capacity over the years due to excessive siltation (Gunnell and Krishnamurthy, 2003; Paranjape et al., 2008); lackadaisical attitude of microlevel institutions managing the tanks, which has mainly stemmed from the growth of private well irrigation in the tank command area resulting in disincentive among farmers to manage these open-access bodies (Balasubramanian and Selvaraj, 2003; Kajisa et al., 2004; Sakthivadivel et al., 2004; ADB, 2006). A few scholars have also highlighted how the hydraulic interdependence between tank storage and aquifer recharge is creating disincentive for farmers to carry out maintenance of tanks, and instead motivating them to privatize these resources by using them as percolation ponds (Sakurai and Palanisami, 2001).

A multivariate analysis by Balasubramaniyam and Bromley (2002) of the factors responsible for tank degradation showed that variables such as encroachments in catchment and water spread area and the increase in canal- and well-irrigation, had significantly increased the degradation of tanks. The increasing importance given to modern irrigation systems, larger reservoirs and river valley projects and the spread of private irrigation wells also have a considerable negative impact on traditional community irrigation systems.

A common view which emerges from the review is that the lack of interest of command area farmers in the management of tanks (Shankari, 1991; Sekar and Palanisami, 2000; Balasubramaniyam and Bromley, 2002), or the erosion of the community management structures, which were responsible for their management, and the subsequent management takeover by the government, had resulted in their decline (PRADAN, 1996; Rao, 1998). Some attributed the loss of community interest in tanks to the advent of groundwater irrigation which gave farmers superior control over irrigation (Dhawan, 1985; Palanisami and Easter, 1991). Some scholars attribute the development approach followed during the British rule, centered on modern large irrigation systems, for the decline of tanks (Paranjape et al., 2008).

However, Mosse (2003, 1999) challenged the long-held view of scholars working on tanks in south India that collapse of community institutions was the major cause of decline of tanks. He contends that even in the past, communities did little investment in the upkeep of tanks. It was the *Zamindars* and kings who, not only built most of the tanks but spent money for their upkeep as well. According to Mosse (2003, 1999), it was the fall of the institution of overlords that led to the decline of the tanks. Shah (2008), while highlighting the cultural and environmental superiority of traditional knowledge that built and managed tanks, as an irrigation technology, argued that tanks have not necessarily produced a democratic social order, either in the past or in the present.

In a nutshell, when viewed together, the work by various scholars seems to suggest that the decline of tanks was due to several events, starting with the takeover of community and Zamindari (private) tanks by the state, which led to institutional erosion, collapse of the system for collection of water charges, and lack of maintenance leading to deterioration of the physical condition of this irrigation infrastructure, and the subsequent gradual loss of community interest in their affairs. Therefore, the factors responsible for decline of tanks were argued to be "institutional."

None of the above theories could fully explain the reasons for degradation of tank systems and decline in tank-irrigated area in South India. In fact, every argument suffers from weaknesses. For instance, if the argument that the advent of groundwater irrigation had really led to the loss of farmer interest in tanks is valid, then it tends to assume that groundwater irrigation is highly equitable and provides farmers from all segments access to and control over well irrigation. This is far from the truth, as pointed out by Kumar (2007). Only a small fraction of the small and marginal farmers in India even today own wells and pump sets (Kumar, 2007), where the situation was much worse in the 1970s and 1980s, when drilling wells was expensive and rural electrification was poor.

As pointed out by Narayanamoorthy (2007), it is the small and marginal farmers, who do not own irrigation wells in the tank commands, who have high stakes in tank irrigation as their livelihood is heavily dependent on it. If this is the case, one cannot explain the poor state of affairs with regard to the condition of tanks in areas such as Kolar (in Karnataka) and Anantapur in Andhra Pradesh (AP), where the poor, small, and marginal farmers do not own wells, and are often dependent on water purchase from well owners engaged in water trading. It is also a notable fact that most of the beneficiaries of tank water for irrigation are small and marginal farmers.

7.3 NEED FOR POSTULATING AN ALTERNATIVE HYPOTHESIS ON TANK DEGRADATION

The dominant theories, advanced by many researchers, harp on about the collapse of traditional tank institutions as the cause of decline of tank irrigation, be it the institutions of overlords, or community management structures. Both theories are also suggestive of a resounding view that performance of tanks as a system is very much within the control of the farmers in the command areas or the institutions which manage them. They inadvertently ignore the fact that these institutions existed in a certain sociopolitical landscape, which is difficult or even impossible to recreate. Such views are based on the assumption that simply cleaning supply channels, or clearing the catchments or repair of the tank embankments (tank bunds), and desilting of the distribution network would yield results in terms of improved storage in tanks and expanded irrigation benefits. They ignore the effect of endogenous and exogenous physical and socioeconomic factors on tank hydrology, and the very impact these effects have on the viability of institutions itself.

First: population growth had a significant demand on irrigation water for crop production in these semiarid regions, which forced farmers to go for alternative sources of irrigation, and the search for this alternative was facilitated by the advent of well irrigation, cheap drilling technology, rural electrification, and subsidized electricity for pumping groundwater. This is quite understandable given the fact that the performance of tanks was subject to high variability in accordance with year-to-year variation in the occurrence of monsoon rains, as these tanks harnessed water only from the local catchments. Furthermore, with a manifold increase in rural population and with an increase in the number of farmers within the limited command of these tanks, the actual area which a single farmer could irrigate using tank water became too low for them to manage their farming enterprise. Well irrigation not only became affordable and in some cases cheaper, it also provided a superior form of irrigation. Yet, as noted by Narayanamoorthy (2007) and Paranjape et al. (2008), for the poor, small, and marginal farmers, the tanks continued to be an important source of irrigation and livelihoods.

Second: the increase in population pressure on private land also meant that farmers had to expand the net area under cultivation, and sometimes this led to encroachment of commons, which formed the original catchments of tanks. Catchment cultivation resulted in a lot of the runoff generated from precipitation being captured through in situ water harvesting for production of rain-fed crops, reducing the inflows into the tanks. Intensive well irrigation on the other hand led to reduced groundwater outflows (base flow) into the upper catchment tanks. Whereas drawdown in the water table resulting from excessive withdrawal of groundwater can potentially lead to greater percolation of water from tanks into the shallow aquifers, further reducing the storage and irrigation potential of the tanks. It is important to mention here that as research has shown in the past, in hard rock areas of peninsular India, the "cone of impression" produced at the bottom of tanks due to percolation of tank water into the shallow aquifer generally stops further percolation of the water (Muralidharan and Athavale, 1998). This peculiar geological and geo-hydrological setting ensured storage of water in the tanks in peninsular India. However, emptying of the aquifer could induce sustained recharge. What is important is that there is no mechanism to control groundwater abstraction by well irrigators in the command and catchments, as legal rights to groundwater are attached to landownership rights.

Reduced irrigation potential of tanks due to the above-cited reasons and the increasing number of tank water users in the command area essentially meant that the contribution tanks could make in the overall livelihood of individual command area farmers, including small and marginal farmers, was too small in comparison to the transaction cost of initiating actions that would improve their performance. In fact, the transaction cost of initiating actions such as removal of encroachments from catchments, and regulating the use of groundwater in the command and catchment would be too high owing to the complicated legal formalities involved, whereas what a farmer could earn in terms of income from crop outputs that can be produced from the use of tank water could be quite insignificant. Whereas, many of the other engineering interventions such as stabilization of tank bund, increasing the capacity of tank through desilting and clearing of tank catchments do not result in incremental benefits that commensurate with the financial investment in most situations. This significantly reduced the incentive among the members of the farming community for self-initiated management actions.

While there are a few tank management activities which can provide substantial private benefits (to the farmers) such as desilting, ie, removal of silt and clay from the tank bed, as the use of it in the field gives direct income benefit to them in terms of higher crop yields for 2–3 years consecutively (Paranjape et al., 2008; Kumar et al., 2011), often the communities lack the wherewithal to take it up. This also explains why the communities come forward to take up tank management activities in situations where there is external support to cover the transaction cost, and the cost of undertaking physical activities.

While the tank potential for meeting the demands for economic activities has been gradually declining in many areas due to the problems described above, improved availability of good-quality water from public water supply schemes within close vicinity of their dwellings has, to a great extent, reduced the dependence of village communities on traditional sources of water such as tanks and ponds for domestic water supplies. Instead, they now depend on these tanks for livestock drinking and washing. This has reduced the village communities' incentive and motivation to protect tank water quality for ensuring potability. Simultaneously, the tanks and ponds have become the natural sink for agricultural runoff containing fertilizer and pesticide residues from upper catchments. This is particularly the case for cascade tanks, wherein the upper catchments of some tanks consist of the command area of the tanks located upstream. The presence of nitrates in agricultural runoff had caused eutrophication of tanks and ponds, more so in the case of high rainfall areas, affecting fish populations in these wetlands.

While it is extremely difficult to say which process has resulted in what outcome, the growing awareness of public health impacts of poor water quality, and recognition that traditional water bodies have increasingly become polluted has also forced the government to think about formal water supply for villages. Here again, it is not just that the communities have lost interest in the protection of tank water quality. There could be situations where there are no alternative sources of water supply in the village. Even in such cases, when the communities desire to protect the tanks from pollution, there is hardly anything which it can do to stop it[1]. Here, both the polluters and those who suffer the damage caused by pollution are in most cases the same people. However, there is a lack of institutional capability to address this issue.

All these factors might have ultimately led to tank degradation in terms of water availability and water quality wherever it has happened. Or in other words, attributing the decline of tanks to one causal factor, ie, erosion of community institutions or collapse of the traditional village institution of "Zamindari" system would be mere oversimplification of a complex physical, socioeconomic, cultural, environmental, and institutional change. Again, what has really led to the decline of tanks in a particular situation is very much case-specific, depending on how the physical, socioeconomic, cultural, environmental, and institutional factors have played out in that particular case. In sum, we would like to make the proposition that technological, socioeconomic, institutional, and cultural changes happening in the societies within which the tanks are embedded have resulted in major physical (hydrological and environmental) changes in tank ecology, which have changed the incentive structures for tank management.

The schemes of minor irrigation departments, NGOs, and donor communities in rehabilitation of tanks have their accent on engineering works, namely, bund stabilization, construction of weirs and repair of sluices, catchment forest clearance and channel desilting, and institutional development, comprising formation of water-user associations and their training for capacity building. They do not pay much attention to the hydrological characteristics of tanks, which are chosen for rehabilitation. Hence they produce poor results, in terms of improving the overall tank performance.

If changes happening in the rural society on the technological, socioeconomic, cultural, and institutional front are bringing about irreversible trends in tank ecology through changes in hydrology and water environment, then programs and projects to improve the condition and performance of tanks that are based on simple engineering interventions such as

[1] This is because pollution is the result of agricultural intensification with more irrigated crops and intensive dairy farming.

desilting, catchment clearing (removal of trees), and supply and distribution channel cleaning (silt removal), are unlikely to lead to any beneficial outcomes. Or in other words, if the rehabilitation program is to be successful there should be sufficient incentive among the potential tank users to take up the rehabilitation work.

7.4 TANK MANAGEMENT PROGRAMME IN ERSTWHILE ANDHRA PRADESH

7.4.1 Characteristics of Tanks in Andhra Pradesh

Out of the total of 18,701 tanks in the erstwhile state of Andhra Pradesh, with water spread area less than 2.5 ha and covering a total wetland area of 610,354 ha, 15,290 are tanks/ponds constituting a total wetland area of 140,000 ha. The rest are man-made reservoirs, water logged areas, etc. The state has three distinct physiographical units. First is the coastal plain to the east extending from the Bay of Bengal to the mountain ranges; Eastern Ghats which form the flank of the coastal plains; and the plateau to the west of the Eastern Ghats. The largest tanks are found in the plateau. The district-wise number of tanks, their wetland area, and the water-spread area during pre- and post-monsoon are given in Table 7.1.

What is most interesting is the fact that waterspread area of these wetlands shrinks drastically after the winter, touching the lowest point during the peak of summer. Table 7.1 shows that summer waterspread area (73,749 ha) is less than half of post-monsoon, ie, November, 2010 (1,632,277 ha). This has major implications for the total water availability of these tanks and the various functions that these tanks can perform in different seasons. Comparison of district-wise data shows that Vizianagaram has the largest number of tanks, followed by Nellore and Medak. But, in terms of average size, tanks/ponds in Cuddappa are the largest, with an area of 25.40 ha.

Another interesting observation is that the ratio of the area irrigated by the tank and the wetland area, which reflect the physical characteristic of tanks, vary widely between districts. For the analysis, we have considered the area irrigated by the tanks in 1970–71, assuming that the real deterioration in tank performance started only later. The wetland area (as estimated through remote sensing imageries) of the tank was taken from the Wetland atlas prepared by the Indian Space Research Organization, Ahmedabad. Since the estimates of wetland area do not consider the tanks with wetland area less than 2.5 ha, it might induce some errors in the estimation of

Table 7.1 Number of tanks and their wetland area in different districts of Andhra Pradesh

Sr. No	Name of district	Total number of tanks/ ponds	Total wetland area (ha)	Average wetland area per tank/ pond (ha)	Total water spread area (ha)	
					Post-monsoon	Pre-monsoon
1	Adilabad	590	7383	12.51	5090	1947
2	Nizamabad	915	19,152	12.51	13,581	7629
3	Karimnagar	692	10,535	20.93	8802	3590
4	Medak	1066	20,116	15.22	16,250	8011
5	Hyderabad	12	107	18.87	71	73
6	Rangareddy	281	3287	8.92	2728	2002
7	Mahbubnagar	340	5424	11.70	3509	2510
8	Nalgonda	601	11,702	15.95	8573	5857
9	Warangal	659	9462	19.47	8433	2367
10	Khammam	558	8044	14.36	5547	2762
11	Srikakulam	460	4678	14.42	3260	1459
12	Vizianagaram	1539	10,567	10.17	7724	2293
13	Visakhapatnam	543	4793	6.87	3409	1341
14	East Godavari	505	4244	8.83	3384	2764
15	West Godavari	334	3191	8.40	2409	1808
16	Krishna	511	6312	9.55	4452	2418
17	Guntur	377	3512	12.35	2678	1939
18	Prakasam	480	7490	9.32	5918	3572
19	Nellore	1088	27,673	15.60	24,890	7151
20	Cuddappa	220	3584	25.43	3069	1458
21	Kurnool	142	1887	16.29	1395	810
22	Anantapur	447	9896	13.29	3131	4135
23	Chittoor	1391	18,638	22.14	17,200	3560
24	Total	15,290	212,244	13.88	163,227	73,749

Source: Authors' own estimates based on National Wetland Atlas, Andhra Pradesh prepared by Indian Space Research Organization, Ahmedabad.

irrigated area–wetland area ratio. The ratio varies from 2.62 in the case of Medak to 23.47 in the case of Srikakulam. Here, Guntur was not considered for the analysis owing to the fact that the tanks in this coastal district receive water from large irrigation schemes. Hyderabad was also not considered as data on tank irrigation in the district was incomplete.

7.4.2 The Tank Management Programme

The Irrigation and Command Area Development (CAD) department of the erstwhile Andhra Pradesh government had undertaken an ambitious program for rehabilitation of tanks coming under the minor irrigation department, which has a design command area of more than 100 acres. The project funded by the World Bank "Andhra Pradesh Community Based Tank Management Project" envisages rehabilitation of 3000 minor irrigation tanks covering 21 districts of the state. The technical or engineering interventions under the rehabilitation program include desilting, jungle clearance, stabilization of bund, sluice repair, waste-weir construction or repair.

As per the project guidelines, the planning, implementation, and post-construction operation and management of the project are to lie with the water users' associations. Software inputs include training of water users' associations for capacity building. To facilitate the planning, implementation, and postimplementation management of various tanks associated with tank rehabilitation, there would be one support organization at a level of a cluster of 5–10 tanks.

The program is demand-driven. The type of activities to be undertaken for rehabilitation of tanks for restoration of irrigation in the command area is to be decided by the water users' association. The funds allocated for this, ie, INR[2] 25,000 per ha of design command area, appears to be quite sufficient at current prices, given the fact that this is for retrofitting of an already-built irrigation infrastructure, and not for building new ones. Again, the program has inbuilt checks and balances for technical soundness and financial management. The plans prepared by the water users' associations are scrutinized by the technical staff of the minor irrigation department after field visits, and they prepare detailed project reports with the estimates for the planned works subsequently. There is a technical manual, which provides the guidelines for design and execution of the physical works and system operation[3].

As the department slightly deviated from its usual operations for this project and added several new activities as the project components, a separate financial manual was prepared exclusively for implementing this project. This document described the procedure for fund flow and arrangement at the state, district, and WUA level, budget preparation, financial management systems at various levels. The manual also provided the procedure for

[2] One USD is equal to about Indian National Rupee (INR) 65.

[3] The manual clearly specifies the guidelines on engineering design, construction, and quality control for the physical components of the work. It also clearly specifies the standards that need to be maintained for each of the tank constituents, be it concrete work, earth work, reinforcement, stone masonry work, revetments, protection of upstream and downstream works, and canal lining.

fund disbursement and accounting system that will be followed in the project. One important financial innovation used in the project was that any expenditure for rehabilitation work below INR 50,000 could be sanctioned by an officer at the rank of Assistant Engineer, and any expenditure in the range of INR 50,000–200,000 could be sanctioned by a Deputy Engineer or Deputy Executive Engineer.

Another notable innovation was with regard to the repayment of water charges being collected from the tank users. Though the cess is collected by the revenue department office at the Mandal level, arrangements are made for quick release of a major share of these funds back to the water users' association through the concerned section office of the irrigation department. This keeps the big incentive for the WUA to motivate the member farmers to clear their water dues in a timely fashion. These funds are to be used for small repair works.

A careful scrutiny of the documents relating to the tank management project shows that a clearcut protocol for picking up tanks for rehabilitation was absent. There was no clearcut policy on the part of the department, which could have been used as the guideline for short-listing the 3000 tanks for rehabilitation, among the many thousands of tanks/ponds in the state. The proposal for the tank management project was prepared on the basis of the feedback and suggestions given by the technical officers (engineers) of the department who are concerned with the management of minor irrigation tanks, from the districts and mandals concerned. One criterion used by the department was to take only those which have a command area of more than 25 ha (ie, 100 acres). The second criterion was the presence of some local initiative by the community for tank management. This, however, was purely based on perception. While this shortcoming could have been overcome through a prudent attempt by the department to involve the local tank communities in this key decision-making, there was no involvement of the tank community whatsoever in selecting the tanks for rehabilitation.

Perhaps, one important consideration which was totally missing in the entire selection process was the hydrological condition of the tanks. Very little attention was paid to the fact that several external factors such as the intensive use of groundwater, catchment encroachment for cultivation are ruining the potential of tanks as reliable sources of water for multiple needs, including irrigation, through their effect on hydrology, which is mostly irreversible. Sufficient efforts are not being made to ensure that the tanks chosen for rehabilitation get sufficient inflows from its catchment. This is also evident from the procedure followed for estimating the catchment runoff. The agency appears to be using the "rational formula," which is a crude method, for estimating the catchment runoff from the total catchment area and a runoff coefficient fixed on the basis of the 75% dependable rainfall. The problem is that for semiarid regions with low to medium rainfalls occurring in an erratic fashion, and for the land cover that exists, the runoff will not be in a linear function of the annual rainfall magnitude, but will increase with rainfall in an exponential fashion. Hence, assuming a fixed value for runoff coefficient will lead to a major error in estimation of runoff.

Ideally, the runoff coefficient will be very low for low rainfall and disproportionately high for high rainfall in semiarid regions with high aridity (Prinz, 2002) causing high variability in runoff (Kumar et al., 2006, 2008). Therefore, the runoff needs to be estimated on the basis of the rainfall–runoff relationship established for the basin or the sub-basin in which the tank catchment falls. This can be done using historical data on streamflows and precipitation for the basin/sub-basin under consideration. Once the rainfall–runoff relation (model) is established, the streamflows for the years (for which records are not available) can be computed on the basis of observed rainfall. From these streamflow values (say for 30–50 years), the dependable runoff (for 75% dependability) can be estimated and used for planning the rehabilitation work. This has to be the most crucial input for deciding on the types of uses from the tank and the cropping pattern. In addition to these, in view of the many changes happening in the rural landscape, including the drainage interception, there is a need for re-assessing the catchment area of the tanks to avoid runoff overestimation.

7.5 RESEARCH OBJECTIVES, APPROACH, METHODOLOGY, AND DATA SOURCES

The specific objectives of the study were as follows: (1) analyze the impact of well development and groundwater intensive use in the tank catchment and commands on tank performance in terms of irrigated area; (2) analyze the impact of catchment cultivation practices on tank performance in terms of area irrigated; (3) identify the physical, socioeconomic, institutional, and environmental factors that result in good overall performance of tanks; and, (4) evolve the criteria for selection of tanks for rehabilitation, and work out the broad management strategies for sustaining and improving the performance of the selected tanks.

The study used analyses of primary data collected from selected performing and non-performing tanks, which were both quantitative and qualitative in nature (for six selected systems) along with secondary data collected from the state minor irrigation department on tank and well irrigation and land use in catchments (on a timescale) at different scales (state, district, block, and individual tanks) for addressing the key research questions. The tanks were selected in such a way that the hydrological and socio-economic environments are not uniform.

A total of six tanks from three districts of the state of Andhra Pradesh were chosen for the study, with two tanks from each district. The locations were chosen in such a way that each one represents a unique situation vis-à-vis the changes in historical performance of tanks. The districts are Vizianagaram, Nizamabad, and Kurnool. Each one falls in a different river basin. Vizianagaram falls in the drainage area of one of the east-flowing rivers, north of Godavari river basin, which is a water-rich river basin. Nizamabad falls in the drainage area of Godavari river basin, which again is a water-rich river basin. The area of Kurnool, which was chosen for the tank case studies, falls in the drainage area of Krishna river basin, which is a water-scarce river basin. Hence, each one represents a hydrological regimen.

In Nizamabad district, as the district-level data show, the area under tank irrigation, in terms of net tank irrigated area, had undergone dramatic reduction, over the past three and a half decades. In Vizianagaram district, the area under tank irrigation had undergone a declining trend during the same period, but not as much as that of Nizamabad. In Kurnool district, no major reduction in area under tank irrigation was observed from the time series data for 35 years from 1970–71 to 2004–05.

Within each district, the two study tanks were selected in such a way that one is in a better condition compared to the other. The judgment about relative performance was made on the basis of the discussions with the officials of the irrigation and CAD department of the government of Andhra Pradesh, who were concerned with the management of these tanks. The criteria used for judging the performance are the condition of the tanks vis-à-vis the physical infrastructure, the area irrigated in the command area against the design command, and the community involvement in their management.

Data were collected from both primary and secondary sources. The primary data were collected both at the tank level and at the farmer level. The tank-level data included: the tank command area; the different uses of tanks and the number of households (HHs) depending on the tank for various uses; area under different crops in different seasons and at different points of time (1970, 1980, 1990, 2000, and at present), and irrigated area under different crops in different seasons and at different points of time; number of wells in the tank command at different points of time; number of wells in tank catchment at different points of time; area under different crops in different seasons in the tank catchment, and irrigated area under different crops in different seasons at different points of time in the tank catchment. They were obtained from the village elders and tank water users' association office bearers. The farmer-level data consist of: (1) area under different crops in different seasons, and at different points of time; (2) irrigated area under different crops in different seasons and at different points of time; (3) changes in cropping pattern and irrigated cropping pattern in drought year, normal year, and wet year; and (4) the current sources of irrigation.

The secondary data comprise the net area irrigated by tank over the period from 1970–71 to 2005–06 in all the districts of AP; the net area irrigated by wells and bore wells over the same time period in all the districts of AP; the area irrigated by different sources, namely, wells, tanks, and canals in the state over the period 1970–71 to 2005–06; and the characteristics of tanks in different districts vis-à-vis their total water spread area, and the net area irrigated.

7.6 RESULTS AND DISCUSSION

7.6.1 Impact of Well Irrigation on Tank Hydrology

It was hypothesized in the beginning of our research that an increase in groundwater use in the catchment and command of the tank would affect the performance of the tanks, through reduction in the inflows into the tank in the form of groundwater outflows or base flows in the streams.

Naturally, base flow contribution to tank inflows would be significant in the case of tanks situated in the upper catchments of river basins, by virtue of the hilly topography and the forested catchments. One example is the tanks located just downstream of the forested catchments of Krishna river basin in Kurnool district. These upper catchment tanks are surrounded on three sides by protected forest area, with good base flows coming in from them. Though in such areas the chances of intensive groundwater use is quite low due to the poor availability of arable land and extremely low groundwater potential in the hard rock formations in the hills, even a small increase in groundwater pumping could alter the hydrological balance.

On the contrary, in the lower catchment tanks, the contribution of base flows to tank inflows could be quite low, and the major contribution of the tank inflows would be from the surface runoff. In these areas, the change in land use is the most important factor altering tank hydrology. The net area under cultivation in these areas is likely to have increased due to increased pressure on land. Here, in this case, it is not an expansion in area under irrigation which is causing changes in the tank hydrology, but increase in area under cultivation. Mostly, it is resulting from encroachment of the common land, which forms the tank catchment, by villagers for cultivation of rainy season crops, which is causing the change.

While normally a part of the rainwater falling in the natural catchments would infiltrate the soil and the remaining water would run-off from the land. While part of this infiltrating water would remain in the soil profile depending on the soil storage capacity, the excess water would percolate down the soil strata and join the groundwater table. The water in the soil profile would eventually evaporate if the soil is barren or else it would support the growth of some natural grass species. But, in any case the rate of evaporation of moisture from the soil profile would be very low. However, once the land is covered by crops, the rate of depletion of moisture from the soil profile would be faster, as the crop would take the water to meet transpiration needs. While this would create more storage space in the soil profile for the incoming precipitation, the presence of vegetation would increase the rate of infiltration. The presence of field bunds would further reduce the downward movement of runoff generated in the field, and the water would percolate down the soil if it is in excess of the moisture deficit in the crop root zone.

In order to understand the dynamics of interaction between tank catchment, groundwater and tank hydrology, we began with the analysis of time series data on net area irrigated by tanks, net area irrigated by open wells and tube (or bore) wells. Historical data on net area irrigated by tanks, open wells and tube wells/bore wells for the period from 1970–71 to 2004–05 are analyzed. The results showing the historical changes in the irrigated area from these three sources are presented in graphical form in Fig. 7.1.

Fig. 7.1 shows that while the area irrigated by open wells increased till the mid-1980s it peaked in 1990–91, and started declining thereafter. Whereas the area under tube wells/bore wells, which tapped the deeper aquifers and geological formations, started increasing exponentially by the early 1990s and continues even today. Overall, the area irrigated by wells increased until 2000–01, when it peaked at 1.95 million hectares.

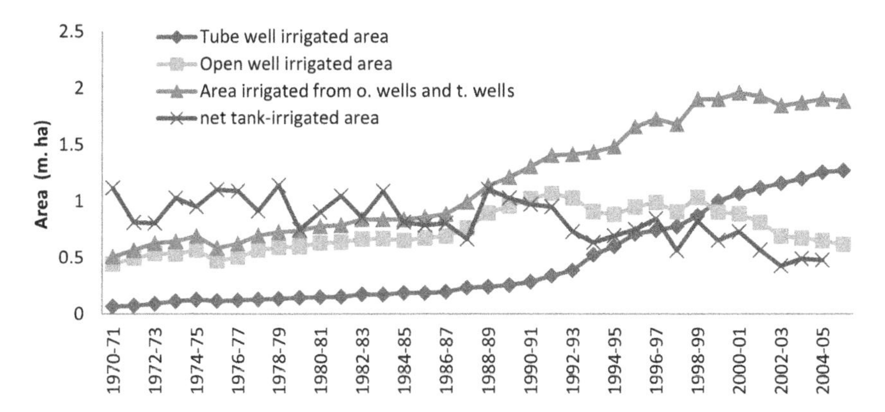

Figure 7.1 Net area irrigated by open wells, tube wells, and tanks, AP.

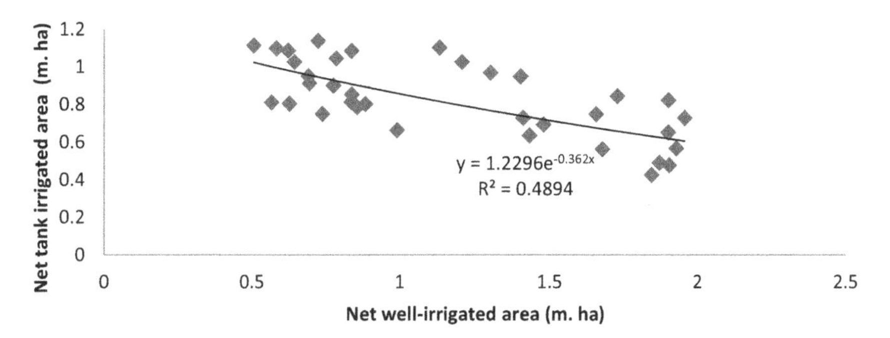

Figure 7.2 Tank-irrigated area versus well-irrigated area.

On the other hand, the (net) area irrigated by tanks started consistently showing declining trends after the late 1980s. Though there have been wide fluctuations in the net area irrigated by tanks between years during the previous years, ie, 1970–71 to 1987–88, such fluctuations could be attributed to interannual variability in rainfall, which will have a direct impact on tank inflows. Regressions run between net tank-irrigated area and net well-irrigated area—the sum of area irrigated by open wells and tube wells—showed a strong inverse relation between the two, with increased well-irrigated area, the area irrigated by tanks reduced linearly ($R^2 = 0.49$) (Fig. 7.2). Furthermore, regression was run with net tank-irrigated area against area irrigated by bore wells. This showed a sharper and stronger relationship (Fig. 7.3). Going by the regression formulas, a unit increase in net tube/bore well-irrigated area resulted in a greater reduction in tank-irrigated areas, as compared to that caused by unit increase in net total

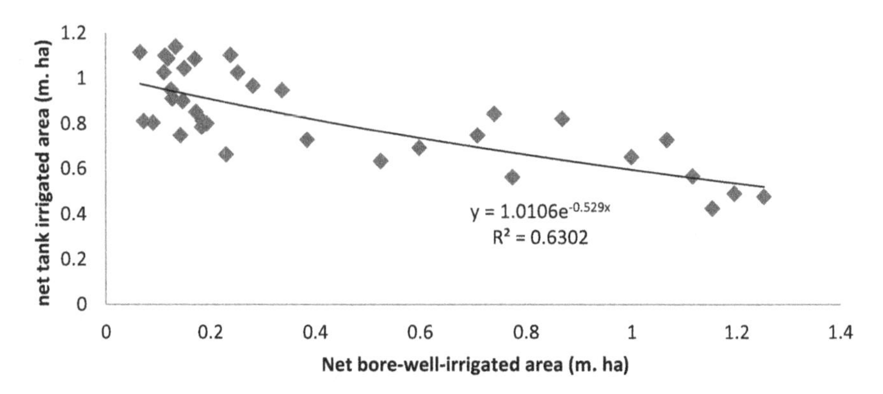

Figure 7.3 Tank-irrigated area versus bore-well-irrigated area.

well-irrigated area. Also, the increase in tube well/bore well irrigation explained the reduction in tank-irrigated area to an extent of 63%, against 49% in the case of total well-irrigated area (Kumar et al., 2011).

The differential trends can be explained in the following way. An increase in well irrigation resulting from increased withdrawal of water from open wells and bore wells suggests greater withdrawal of groundwater. This would normally affect the base flows into tanks, and also the percolation of stored water in the tanks into the formations underlying it, depending on where the tanks are located within the basin. But, at the same time, a consistent increase in area under well irrigation till around the year 1991–92 can also be suggestive of the fact that unsustainable levels of abstraction of the shallow aquifer were not reached till that point of time. Under such circumstances, the effect of groundwater withdrawal on tank inflows will be less. This explains the milder slope of the curve representing decline in tank irrigation with the increase in well irrigation.

Now, as regards the lower regression value, other than well irrigation, rainfall can also be an important factor which explains the changes in tank-irrigated area. In good rainfall years, the tank inflows, including those from groundwater outflows or base flows, could also be high, especially when the open well irrigation is still dominant. Nevertheless, there would be a rise in pumping above the normal year values as a result of better replenishment of groundwater. In contrast, in low rainfall years, the runoff and base flows could reduce, along with groundwater recharge. So, in such years, the groundwater irrigation from open wells along with tank irrigation would be less than that of normal year values. This would upset the normal trend. Such trends in well irrigation, which is characteristic of hard rock regions with poor static groundwater resources, are visible in Andhra Pradesh (see Fig. 7.1).

At the same time, an increase in tube/bore well irrigation (from 0.98 million ha in 1990–91 to 1.27 million ha in 2004–05 by around 0.99 million ha), which is also accompanied by a reduction in open well irrigation, indicates the dewatering of shallow aquifers, and the pumping of water from the deeper strata. This not only means the chances for base flow contributing to streams flows into tanks are almost absent, but the possibility of tanks losing their storage into the dewatered aquifers would also be very high. Again under such geo-hydrological environments of deep dewatered zones, the response of shallow and deep aquifers to incident rainfall would be slower as compared to a situation where the shallow aquifer is saturated owing to the time taken for water to move from the top soil to the water table and the amount of water lost while percolating. Here, unlike in the earlier case, good rainfall may not result in proportional improvement in groundwater recharge and base flows, while tank inflows and tank-irrigated area could increase due to increase in runoff. Therefore, rainfall will have a lesser influence on groundwater–tank interactions. This increases the negative effect of tube well/bore well irrigation on tank performance.

While intensive well irrigation explains the decline in tank irrigation to the extent of 48–63% (see Figs. 7.2 and 7.3), clearly there are other factors which cause a decline in tank-irrigated area. We have hypothesized that change in land use would be another important factor, which causes a reduction in tank inflows and tank-irrigated area. However, we can test this hypothesis only using primary data collected from the field, as the secondary data on land use are not available for individual tank catchments, but also mandals and blocks and districts.

7.6.2 Differential Impacts of Well Irrigation on Tanks Across Different Regions

The foregoing analyses do not, however, suggest that the historical tank performance has been uniform across the board. In fact, analysis of tank-irrigated area, well-irrigated area and bore-well-irrigated area for different districts shows distinctly different trends. For instance, in Nizamabad district, the decline in tank-irrigated area was very sharp; an inverse linear relationship exists between tank-irrigated area and well-irrigated area; and the regression value is very high ($R^2 = 0.55$ for both total well-irrigated area and bore-well-irrigated area). For every one hectare increase in well-irrigated area, the tank-irrigated area declined by 0.25 ha. The relationship was quite weak in the case of Vizianagaram district. Interestingly, no relationship seems to exist between net tank-irrigated area and net well-irrigated area (and also net bore-well-irrigated area) in the case of Kurnool district. In fact, the net area under tank irrigation did not show any consistent decline in the district, which has large tracts of forest land and many tanks, which have forested catchments.

Further analysis was carried out to examine whether any relationship exists between tank characteristics, defined by the ratio of the area irrigated by the tank and its wetland area, and the degree of reduction in irrigated area over time, which these tanks have undergone. The analysis was carried out using district as the unit. The percentage reduction in irrigated area was estimated for each district by taking the ratio of the reduction in area during 1970–71 to 2004–05 and dividing it by the area irrigated in 1970–71. Here, we assume that the area irrigated by the tank in 1970–71 reflects the best irrigation performance for the tanks in all the districts. Analysis showed an inverse (logarithmic) relationship between the "irrigation–wetland area ratio" and the reduction in area which irrigation from the tank has undergone, defined in percentage terms. The greater the value of the wetland–irrigation ratio, the lower the reduction in area irrigated over time. Frequency analysis showed that tanks with a wetland–irrigation ratio in the range of 2.0–5.0 experienced an average reduction in irrigated area of 73.5%. Against this, those tanks with a wetland–irrigation ratio exceeding 5.0 (between 6.26 and 23.47) experienced an average reduction in the irrigated area of only 48.70%. Many districts in this category showed less than a 25% reduction in irrigated area. These analyses indicate that the tanks with low irrigation–wetland ratio are likely to deteriorate much faster than those with high irrigation–wetland ratio.

7.6.3 Impact of Catchment Land Use on Tank Performance: Analysis of Field Data

The impact of changes in land use in the catchment on the performance of tanks was analyzed by comparing temporal changes in total area under different crops and tank-irrigated area under different crops in different seasons in the command area against the temporal changes in types and density of groundwater abstraction structures and cropping in the catchment. The impact of land-use changes on tank performance was also analyzed by comparing the performance of two tanks in terms of percentage area under irrigation and irrigated area ratio at a given point of time against the characteristics of land use in the catchment of the tanks such as well density and area under cropping at that point of time.

The analysis was done for both tank-level data and individual farmer-level data. The tank-level data included: (1) area under different crops in different seasons in the command area and catchment area; (2) area of different crops irrigated by tank water in different seasons in the command area; and, (3) density of wells in the command area and catchment area.

The summary of analysis for the tank-level data is provided in Table 7.2. It provides the data on performance of the six tanks in terms of gross cropped area, gross irrigated area, irrigated area as a fraction of the cropped area, and irrigated area ratio for the command area, against the well density in command, well density in catchment, and gross cropped area in the catchment. It covers the tanks which are performing well, as well as tanks which are not performing so well from all the three districts. As regards tank performance, two types of comparison were possible. (1) The difference in temporal performance of tanks located in different districts. (2) The difference in historic performance of tanks from the same location (district). This is compared against the physical and socioeconomic features of the tanks under study to understand the reasons for differential performance.

The analysis shows that tanks in Kurnool are the best tanks in terms of the overall condition as it had sufficient storage even in the beginning of summer season. The catchment area of the tank is covered by reserve forests and as such has no cultivation or wells. Though the irrigated area ratio (ratio of gross irrigated area against the command area) is small for the first tank (ranging from 0.53 in 1980 to 0.44 at present), a large amount of water from the tank is diverted for domestic water supplies. The survey showed that in addition to irrigation, the 300 families from the village also use tank water for livestock drinking. Second, the command area farmers of the first tank (relatively better condition) grow water-intensive crops such as chilli along with paddy and jowar, whereas farmers in other district tanks grow low water-consuming, short-duration pulses such as green gram and black gram, which largely use residual soil moisture from the harvested paddy fields. Further, unlike in the case of both the tanks in Nizamabad and one of the tanks in Vizianagaram, there are fewer wells in the command of the tanks surveyed in Kurnool.

As Table 7.3 shows, all the 26 farmers surveyed in the command were found to be using only tank water in Parumanchala tank. In the case of Padmaraja tank, only two of the 33 farmers surveyed were using well water. Hence, it can be very well assumed that in the case of Kurnool tanks,

Table 7.2 Tank performance versus tank characteristics

Name of district	Tank no.	Good/ Degraded	Name of the tank	Command area (ha)	Gross cropped area in the command (Ha)					Gross irrigated area in the command (ha)				
					1970	1980	1990	2000	Current	1970	1980	1990	2000	Current
Kurnool (A)	1	Good	Parumanchala	607.0		324.0	301.7	313.9	269.3		324.0	301.7	313.9	269.3
	2	Degraded	Padmaraja tank	344.3			368.6	405.0	328.1			368.6	405.0	328.1
Nizamabad (B)	1	Good	Gundla tank	122.31	121.41	141.75	141.75	170.10	202.50	121.41	141.75	141.75	170.10	202.50
	2	Degraded	Jukkul tank	63.99	64.80	70.47	70.47	70.88	95.18	64.80	70.47	70.47	70.88	95.18
Vizianagaram (C)	1	Good	Pedda tank	303.51	526.50	567.00	567.00	526.50	546.75	413.10	405.00	425.25	425.25	441.45
	2	Degraded	Nalla tank	127.48	255.15	255.15	255.15	255.15	145.80	255.15	255.15	255.15	255.15	137.70

Tank Code	Percentage area irrigated					Irrigated area ratio for the year					Density of wells in the command area					Density of wells in the catchment area				
	1970	1980	1990	2000	Current	1970	1980	1990	2000	Current	1970	1980	1990	2000	Current	1970	1980	1990	2000	Current
A1		1.0	1.0	1.0	1.0		0.53	0.50	0.52	0.44		0.008	0.008	0.008	0.008		0.04	0.04	0.09	0.09
A2			1.0	1.0	1.0			1.07	1.18	0.95			0.049	0.049	0.049	There are no wells in the catchment area				
B1	1.00	1.00	1.00	1.00	1.00	0.99	1.16	1.16	1.39	1.66	2.45	2.45	2.58	2.62	2.70	There are no wells in the catchment area				
B2	1.00	1.00	1.00	1.00	1.00	1.01	1.10	1.10	1.11	1.49		1.56	2.11	2.19	2.31	0.64	0.84	1.01	1.63	1.88
C1	0.78	0.71	0.75	0.81	0.81	1.36	1.36	1.40	1.40	1.45	0.21	0.21	0.20	0.16	0.20	0.124	0.124	0.082	0.049	
C2	1.00	1.00	1.00	1.00	0.94	2.0	2.0	2.0	2.0	1.1	There are no wells in the command area					There are no wells in the catchment area				

Tank Code	Gross cropped area in the catchment					Catchment	Percentage area under cropping in the catchment					Major tank uses			
	1970	1980	1990	2000	Current	Area (ha)	1970	1980	1990	2000	Current	Irrigation	Domestic water supply	Fisheries	Livestock
A1	There is no crop cultivation in the catchment					137.7						300	300	8	300
A2	There is no crop cultivation in the catchment					176.58						250	300	10	300
B1			40.47	80.94	202.34	243.00	0.00	0.00	0.17	0.33	0.83	300		100	1000
B2	38.45	38.45	44.52	40.47	182.11	202.50	0.19	0.19	0.22	0.20	0.90	150		30	300
C1	141.64	121.41	161.87	162.00	NA	121.50	1.17	1.00	1.33	1.33		1000		100	
C2	No data available				188.18	510.00	0.00	0.00	0.00	0.00	0.37	200		50	

Source: Authors' own analysis using primary data from tank water users.

Table 7.3 Percentage of sample farmers using different sources of irrigation in the tank commands

	No. of farmers using different sources of irrigation in the command of					
	Nizamabad		Kurnool		Vizianagaram	
Name of tank	Gundla tank	Jukkul Tank	Parumanchala Tank	Padmaraja Tank	Pedda Tank	Nalla Tank
Tank	29	12	26	28	20	20
Open well		6		2	1	
Tank and open well		8			2	
Bore well	5	3		1	3	
Tank and bore well	7	3			1	
Rain-fed		4		2	1	
Tank and rain	1	11				
Open well and bore well		1				
Open well and rain		1				
Total	42	49	26	33	28	20
Percentage farmers using different sources for irrigation						
Tank	69.0	24.5	100.0	84.8	71.4	100.0
Open well		12.2		6.1	3.6	
Tank and open well		16.3			7.1	
Bore well	11.9	6.1		3.0	10.7	
Tank and bore well	16.7	6.1			3.6	
Rain-fed		8.2		6.1	3.6	
Tank and rain	2.4	22.4				
Open well and bore well		2.0				
Open well and rain		2.0				

Source: Authors' own analysis using primary data from tank water users.

a remarkable share of the irrigation in the command is from tank water. In the case of Nizamabad, the well density in the command area of the degraded Jukkul tank has been increasing over time, and is very high now. Twenty-two out of the 49 farmers surveyed from the tank command reported using well water in conjunction with tank water, while 12 of them used only well water. Hence, most of the irrigation reported in the command area must be from wells only. Therefore, though the irrigated area ratio for the tank command has been increasing, comparison of tank performance based on those data alone will be highly misleading. Interestingly, though Nalla tank is highly degraded, in the absence of wells, the farmers in the command only depend on tank water. Hence, the bad condition of the tank is reflected in the irrigated area, which has reduced over time.

Comparison between two tanks in the same location brings out the effect of catchment land-use changes in a better way. In the case of Gundla tank in Nizamabad and Pedda tank in Vizianagaram, which were perceived as in good condition by the local people, the irrigated area ratio has increased over time. For instance, in the case of Gundla tank, it increased from 0.99 in 1970 to 1.66 at present, whereas in the case of Pedda tank, it increased from 1.36 in 1970 to 1.45 at present. This is quite contrary to what was found in the districts in terms of historical performance of tanks. In both the districts, the tank-irrigated area declined, while the decline was very drastic in the case of Nizamabad.

What makes these tanks distinct is the fact that there is no significant groundwater use in their catchments. While there are no wells in the catchment area of Gundla tank, there are very few wells in the catchment of Pedda tank (one well per 20 ha of catchment). But, this increase in irrigated area cannot be attributed to the good condition of tanks alone. The current groundwater use in the command area is significant for winter and summer crops, though these are low water-consuming crops. As seen during the field work in Vizianagaram, almost all farmers in the command area of Pedda tank have bore wells with electricity connections. Wells are the main source of water for irrigation of winter crops (maize) and summer crops. Even during the month of summer, a significant portion of the command area is under sesame, which is a short-duration crop.

Contrary to these, in the case of Jukkul tank (degraded tank of Nizamabad), there are around 37 wells per 20 ha of the catchment area, as of today, which actually experienced an increase from around 13 wells per 20 ha in 1970 to 20 wells per 20 ha in 1990. Furthermore, the area under cultivation in the catchment as of today is 90%, and it increased dramatically from 38.45 to 182.1 ha over a period of 40 years.

Though there are no wells in the catchment of Nalla tank (degraded tank of Vizianagaram) in Vizianagaram district, rain-fed cultivation in the catchment has drastically increased over time—to 188 ha at present, covering 37% of the catchment area. Probably due to this, the irrigated area ratio for the tank declined drastically from 2.0 to 1.1 during the period from 2000 to 2011. This means that the actual performance of tanks in terms of area served by it is much lower than the reported (irrigated) area in the tank command and in reality, it might have declined.

7.6.4 Agricultural Activities in the Command and Catchment of Tanks

The changes in crop production in both the commands and catchments of the tanks are analyzed using primary data collected from 25 farmers located in the command and catchments from each tank location, with a total of roughly 150 farmers from the six tanks selected for investigation. While the changes in cropping pattern and cropped area in the catchment are likely to influence the tank hydrology, the changes occurring in the cropping pattern and cropped area in the command area are a reflection of the changes in tank hydrology, and other farming-related externalities which include the change in access to water from underground and other sources. As discussed in the first section, the scenario vis-à-vis access to groundwater had dramatically changed in Andhra Pradesh after the 1980s with the advent of energized pump sets for extracting water from open wells, and bore wells that are able to tap water from deeper strata. If market conditions remain the same, reduced inflows into the tanks can force farmers to reduce the area under water-intensive crops like paddy and sugarcane and shift to low water-consuming crops like pulses. However, improved access well water, through energized pump sets and drilling technologies can enable farmers to intensify their cropping and go for irrigated crops during winter and summer seasons. The results of the analysis are presented district-wise and tank-wise below.

7.6.4.1 Tank 1: Parumanchala Tank, Parumanchala Village, Nandikotkur Mandal, Kurnool District

Table 7.4 shows the cropped area and cropping pattern of Parumanchala tank command for the period from 1970 to 2011 (current). The analysis shows that there has been a remarkable increase in the cropped area in the command during the past 40 years, with most of the expansion occurring in the first decade, ie, 1970–80. During this period, the cropping intensity also increased substantially, with crops being cultivated during winter and summer. Thereafter, it had hovered around 155–165 ha. Even during 2000–11, the change in gross cropped area was negligible. While the area under kharif paddy did not show any major change during 1970–2000, the percentage area under kharif paddy drastically declined (from 69% in

Table 7.4 Gross cropped and irrigated area (acre) of sample farmers in Parumanchala tank command

Season	Crop	Area under the crop in				
		1970	1980	1990	2000	Last normal year
Kharif	Paddy	30.2	30.2	30.2	30.2	8.0
	Maize			5.0		7.5
	Jowar	5.0	21.5	23.0	29.5	57.7
	Cotton	8.5	2.0	2.0	2.0	2.0
	Chilli		46.3	35.3	63.7	45.3
	Groundnut		7.0	7.0	7.0	12.0
	Tobacco		15.0	15.0	15.0	15.1
	Sunflower		7.5	17.5	5.0	5.0
	Pigeon pea		10.0	10.0	10.1	15.1
	Black gram			6.0		
Rabi	Black gram		2.0	2.0	2.0	
Summer	Jowar		1.5			
	Tobacco		15.0			
	Chilli		15.0			
	Pigeon pea		10.0			
Total		43.7	182.9	152.9	164.5	167.7
Percentage area of crops in Parumanchala tank command						
Kharif	Paddy	69.1	16.5	19.7	18.4	4.8
	Maize			3.3		4.5
	Jowar	11.4	11.8	15.0	17.9	34.4
	Cotton	19.4	1.1	1.3	1.2	1.2
	Chilli		25.3	23.0	38.7	27.0
	Groundnut		3.8	4.6	4.3	7.2
	Tobacco		8.2	9.8	9.1	9.0
	Sunflower		4.1	11.4	3.0	3.0
	Pigeon pea		5.5	6.5	6.1	9.0
	Black gram			3.9		
Rabi	Black gram		1.1	1.3	1.2	
Summer	Jowar		0.8			
	Tobacco		8.2			
	Chilli		8.2			
	Pigeon pea		5.5			

Note: The gross cropped and irrigated areas in Parumanchala tank command are one and the same.
Source: Authors' own analysis using primary data.

Table 7.5 Gross cropped and irrigated area (acre) of sample farmers in Parumanchala tank catchment area

Name of season	Name of Crop	1980		1990		2000		Last Normal Year	
		Cultivated	Irrigated	Cultivated	Irrigated	Cultivated	Irrigated	Cultivated	Irrigated
Kharif	Groundnut	24	24			24	24		
	Red gram			24	24			14	14
	Tobacco							10	10

Source: Authors' own analysis based on primary data.

1970 to 4.8% in 2011), as new crops were introduced in the command. These new crops include chilli, groundnut, tobacco, summer jowar, cotton, and pulses. Chilli occupies nearly 45% of the cropped area of the sample farmers, and is second only to jowar, which is the major kharif crop in the tank command today.

The corresponding figures for the tank catchment are shown in Table 7.5. The data show that cropping is practiced only during kharif season, with groundnut and red gram being the major crops. These crops require much less irrigation water as compared to paddy. Though these are also reported to be irrigated, the availability of rains means that they require very little irrigation even during dry spells.

7.6.4.2 Tank 2: Padmaraja Tank of Indireswaram Village, Atmakur Mandal, Kurnool District

Table 7.6 shows the cropped area, irrigated area, and cropping pattern in the command area of Padmaraja tank, which is considered to be a well-performing tank in the area for the period from 1980 to 2011. Since all the crops are irrigated in almost all situations, as one can see, the cropped area figures and irrigated area figures are more or less the same, except in 1990. As one can see, there has been no significant change in the cropped area in the tank command since 1980, except a 15% increase during 1990–2000 as compared to the base year of 1980. Thereafter, the area under cropping and irrigation reduced slightly. The tank command also did not witness much change in the cropping pattern. The area under paddy hovered around 69–85% during the past 3 years. Unlike in the case of Parumanchala tank, where the percentage area under paddy reduced drastically after 1970, in this case, it went up slightly during 2000–11. Nevertheless, no major cropping was reported during the winter season, while at least four crops, namely, cotton, sunflower, chilli, and jowar, were grown by farmers during the season during 1980–2000.

As regards tank catchment, no crops were grown there, as it is under reserve forest.

Table 7.6 Gross cropped and irrigated area (acre) of sample farmers in Padmaraja tank

Name of season	Name of crop	1980		1990		2000		Last normal year	
		Cropped	Irrigated	Cropped	Irrigated	Cropped	Irrigated	Cropped	Irrigated
Kharif	Paddy	119.8	119.8	92.0	92.0	145.3	145.3	135.8	135.8
	Cotton	3.0	3.0	6.0	6.0				
	Sunflower	22.5	22.5	9.3	3.0	11.0	11.0	4.0	4.0
	Maize	2.0	2.0	2.0	2.0	2.0	2.0		
	Jowar			22.0	22.0				
	Jute					3.0	3.0	3.0	3.0
	Gooseberry							17.0	17.0
Rabi	Cotton	2.0	2.0	2.0	2.0	2.0	2.0		
	Sunflower	3.5	3.5	3.5	3.5				
	Chilli	1.0	1.0	1.0	1.0	21.5	21.5		
	Jowar	2.0	2.0	2.0	2.0	2.0	2.0		
	Watermelon					1.0	1.0	1.0	1.0
Total		155.8	155.8	139.8	133.5	187.8	187.8	160.8	160.8
Percentage area of crops cultivated and irrigated in the Indireswaram tank command									
Kharif	Paddy	76.9	76.9	65.8	68.9	77.4	77.4	84.5	84.5
	Cotton	1.9	1.9	4.3	4.5				
	Sunflower	14.4	14.4	6.7	2.2	5.9	5.9	2.5	2.5
	Maize	1.3	1.3	1.4	1.5	1.1	1.1		
	Jowar			15.7	16.5				
	Jute					1.6	1.6	1.9	1.9
	Gooseberry							10.6	10.6
Rabi	Cotton	1.3	1.3	1.4	1.5	1.1	1.1		
	Sunflower	2.2	2.2	2.5	2.6				
	Chilly	0.6	0.6	0.7	0.7	11.4	11.4		
	Jowar	1.3	1.3	1.4	1.5	1.1	1.1		
	Watermelon					0.5	0.5	0.6	0.6

Source: Authors' own analysis using primary data.

7.6.4.3 Tank 3: Jukkul Tank, Bhavanipet Village, Machareddy Mandal, Nizamabad District

A quick review of the data on gross cropped area and cropping pattern shows that there has been a significant reduction in the gross cropped area since 1970—with the gross cropped area declining from nearly 60 acres in 1970 to around 36 acres in the last normal rainfall year.

As regards the cropping pattern, no pattern seems to be emerging, except that during dry years, farmers reduce the area under both kharif and winter paddy. Another important observation is that sugarcane was introduced as a major crop during 1980s, but had almost disappeared by 2000. This could be attributed to the introduction of energized wells in the area, and increase in command area farmers' access to well water for irrigation. Currently, the farmers in the command do not grow sugarcane. Overall, reduction in cropped area is a strong indication of water shortages faced by tank irrigators.

Since Jukkul tank is a degraded tank, it is important to examine the changes in farming systems in the catchment area. The data show that the area under cropping has increased in the command area, from 11.7 acres in 1970–21.20 acres in the last normal year (2011). An interesting observation is that sugarcane was introduced as a major crop as the data for 1990 indicate, but farmers did not continue this crop. Paddy continues to be the dominant crop in the catchment, occupying around 80% of the total cropped area, though in 2000 maize was the major crop, occupying around 75% of the gross cropped area.

As regards the irrigation in the command, there has been an overall reduction in the area irrigated by tank—from 57 acres in 1970 to 34 acres last normal year, ie, 2011, though maximum shrinkage in irrigated cropped area was reported for 1990 (by nearly two-thirds). Thereafter, the irrigated area improved to become 45 acres in 2000 and then declined. In line with data on cropping pattern, the maximum area under irrigated sugarcane was reported in 1990 (around 28 acres), occupying around 75% of the total irrigated area in the command. A useful observation vis-à-vis irrigated cropping pattern is of maize becoming a dominant crop during drought years. During the last drought, it was reported to have occupied around 86% of the total area under irrigation.

As regards the catchment land use, the irrigated area is more or less the same as that of cropped area, meaning all the crops cultivated in the catchment are also irrigated.

7.6.4.4 Tank 4: Gundla Tank, Domakonda Village and Mandal, Nizamabad District

Gundla tank is a better-performing tank when compared to Jukkul tank. The outcomes of the analysis of data on cropped area and irrigated area of different crops in the tank command are presented in Table 7.7. In the case of Gundla tank, no notable and consistent reduction or increase in area reported by the farmers was seen over the years. What is more important is the fact that paddy remained as the most dominant crop in the command, raised during both kharif and winter season and occupying more than 80% of the cropped area.

7.6.4.5 Tank 5: Pedda Tank, Rellivalasa, Vizianagaram District

In the case of Vizianagaram tanks, no cultivation in the catchment was reported by farmers though the village survey showed some cultivation in the catchment area. As per the tank-level data obtained from the village tank users' association, while no data on cropping are available for the present situation in the case of Pedda tank, no data on historical cropping (1970–2000) are available for Nalla tank. Though a lot of discrepancy is observed vis-à-vis data on land use in the catchment, the fact remains that there are no wells in the catchment area, which is suggestive of low intensity of land use there in terms of agriculture. Hence the outputs were generated only for cropped area and irrigated area in the tank command.

In the case of Pedda tank, which is considered to be one in good condition, the area under cropping had consistently increased from 1970 to 2000.

Table 7.7 Area under different crops (acre) and irrigation of sample farmers in Gundla tank command

Name of season	Name of crop	Area under the crop in					
		1970	1980	1990	2000	Last normal year	Last drought year
Kharif	Paddy	51.6	62.9	6.9	64.4	61.9	57.6
	Maize					5.0	9.0
Rabi	Paddy	48.2	62.9	57.0	41.5	45.9	26.0
	Maize			2.0	2.0	5.0	10.0
Total		99.8	125.9	66.0	107.9	117.8	102.6
Kharif	Paddy	51.7	50.0	10.5	59.7	52.6	56.1
	Maize					4.2	8.8
Rabi	Paddy	48.3	50.0	86.4	38.5	38.9	25.4
	Maize			3.0	1.9	4.2	9.8

Source: Authors' own analysis using primary data.

But, later the total cropped area of sample farmers declined to around 108 acres in 2011, the last normal year. The increase in cropped area was also accompanied by an increase in area under paddy, winter groundnut, and summer maize. As regards the cropping pattern, the percentage area under paddy, groundnut, and maize did not fluctuate much over the years. The area under irrigation was also found to be same as that under cropping as all the crops were irrigated. Currently, farmers are found to be growing sesame during later summer, which lasts till the end of July.

7.6.4.6 Tank 6: Nalla Tank of Pinavemali, Vizianagaram Mandal and District

In the case of Nalla tank, the total cropped area of sample farmers surveyed in the command consistently decreased from 153.7 acres in 1970 to as low as 91 acres in 2000, and then increased to 165 acres in the last normal year (2011). However, a substantial portion of this increase in the area came from pulses such as green gram and black gram, which are very low water-consuming, short-duration crops. The area under green gram and black gram went up by 52 acres. Though area under kharif paddy, which receives only supplementary irrigation, was very high during the last normal year (a rise of 38 acres from year 2000 figures), there was a reduction in area under winter maize, which is fully irrigated. The gross cropped area was, however, down to 116 acres in the last drought year. The percentage area under paddy did not change significantly during the 40-year period from 1970 to 2011, though it was highest during 1990. The irrigated area in the Nalla tank command is a little lower than the cropped area, with a small portion paddy left unirrigated. Paddy occupies nearly 45% of the cropped area.

7.6.5 Performance of Tanks Against Their Physical and Socioeconomic Attributes

To analyze the effect of groundwater irrigation and crop cultivation in the catchment on tank performance, we examined irrigated area ratio for the tanks in question. However, our analysis shows that this can often be misleading due to two reasons. First, the cropping pattern changes drastically from tank to tank, and often is a function of the tank hydrology itself. The farmers in the command area of tanks which receive sufficient inflows tend to grow highly water-intensive crops such as chilli, sugarcane, and sunflower, whereas those in the command area of tanks facing water shortage tend to grow low water-consuming crops such as pulses after kharif paddy. From perennially water-rich tanks, water is also used for domestic water supplies in the neighboring villages and small towns, apart from livestock drinking.

Second, the presence of wells in the tank command alters the scenario of the irrigated area in the tank command. Even in instances where the tanks are not able to serve the command area farmers because of reduced inflows from the catchments, the wells in the command area meet the crop water requirement. However, under no circumstances does the presence of wells in the command influence the farmers not to use water from the tanks. Instead, it is the inability to get sufficient water from tanks for irrigating crops that motivates them to go for well irrigation. Therefore, high density of wells in the command area should be treated as a sign of low dependability of the tank system as a source of water for irrigation and other uses in the command.

When the analysis looked at the variations in cropping pattern in the command area, the uses of tank water other than for irrigation (like domestic water supply to towns and villages and livestock drinking), and the effect of wells in the tank command on irrigation performance to do the comparative performance of tanks, the following becomes clear. The performance of those with high density of wells in the catchment area and high land-use intensity are likely to decline drastically over time, whereas those without much interception of their catchments through farming and groundwater withdrawal sustain their performance, without any threat to the hydrological integrity. The best-performing tanks will be those which have no cultivation in the catchment (which also implies that there are no wells), and the second best are those which have no wells in the catchment but have rain-fed cultivation. As a result, tanks which have their upper catchments located in forests are the most ideal in terms of performance.

7.7 FINDINGS

Several researchers have inquired into the causes for decline of tank irrigation in south India. The range of factors to which they have attributed the "decline," are largely social and institutional. The most dominant of them are the lack of incentive among the command area farmers and collapse of traditional management institutions, including community management structures and institutions of overlords (Zamindars) which took care of their upkeep, increase in groundwater irrigation and the consequent reduction of interest among farmers in tanks, interception of supply channels, and lack of adequate attention paid to regular maintenance.

We propose an alternative hypothesis that excessive groundwater draft characterized by groundwater irrigation in the tank catchment and commands, and land-use changes in the catchment in the form of intensive crop cultivation resulted in reduced tank inflows, causing a decline in the area irrigated by tanks. These hypotheses were tested using: (1) secondary data available from government agencies at the district level on area irrigated by tanks and wells; and (2) primary data collected from tank communities at the local level on changes in groundwater irrigation in catchments and commands, and changes in catchment land use.

Groundwater irrigation has been growing steadily in Andhra Pradesh since the early 1970s till the end of the last century, as indicated by the figures of net total well-irrigated area. This has been evident through an increase in the number of deep bore wells and energized pump sets for open wells and bore wells. As our analysis shows, the net well-irrigated area began to "plateau" after 2000–01. Thereafter, as pointed out by Kumar et al. (2011), the increase in number of wells had not resulted in an increase in gross well-irrigated area either.

Analysis of the dynamics of the interaction between groundwater and tanks at the level of the state and districts showed that increased groundwater draft adversely affected the performance of tanks, as indicated by the strong correlation between well- and bore-well-irrigated area and net area irrigated by tanks at the state level for AP, and also for many districts of the states. Therefore, it could be safely argued that much of the expansion in well-irrigated area happened at the cost of tank irrigation. Nevertheless, the effect of well irrigation on tank performance has not been uniform. While in many districts, the decline in tank performance in terms of "net area irrigated by tanks" in response to increase in "net well-irrigated area" has been sharp, in some districts, there has not been much reduction in the net tank-irrigated area, in spite of a remarkable increase in well-irrigated area.

In lieu of the adverse impact of well irrigation on tank performance, it could be stated that though the net increase in well-irrigated area in the state has been a remarkable 1.4 million ha (net), the overall contribution of wells to expansion of irrigation in the state will be much less, if one takes into account the fact that the reduction in net tank-irrigated area is around 0.60 million ha, ie, from 1.11 million ha to 0.47 million ha. However, the reduction in tank-irrigated area cannot be fully attributed to groundwater overextraction, and part of the reduction might have been caused by the change in land use in the tank catchments.

Analysis carried out to examine the relationship that exists between tank characteristics, and the degree of reduction in irrigated area over time showed an inverse (logarithmic) relationship between the "irrigation–wetland area ratio" and percentage reduction in irrigated area of the tank. The greater the value of the wetland–irrigation ratio, the lower the reduction in area irrigated over time. Frequency analysis showed that tanks with a wetland–irrigation ratio in the range of 2.0–5.0 experienced an average reduction in irrigated area of 73.5%. Against this, those tanks with a wetland–irrigation ratio exceeding 5.0 (between 6.26 and 23.47) experienced an average reduction in the irrigated area of only 48.7%. Many districts in this category showed less than a 25% reduction in irrigated area. These analyses indicate that the tanks with low irrigation–wetland ratios are likely to deteriorate much faster than those with high irrigation–wetland ratios.

As regards the impact of catchment land use on tanks, some are positive, some are negative. The first type of change is in the area under cultivation in the catchment. The catchments of tanks are generally public land with government forests, pasture land, and revenue wasteland. Barring the reserve forests, these catchments are increasingly being encroached upon by individual villagers for cultivation. One factor which triggered intensive land use in the catchment is the access to well irrigation. With water available from wells for supplementary irrigation, farmers are able to take up cultivation of kharif crops even in the driest regions of the country. Hence, intensive groundwater irrigation in the catchment had a double impact on tank hydrology, first by affecting the groundwater outflows into streams and then by affecting the runoff from the catchment entering the tank reservoir.

Cultivation alters the catchment hydrology by reducing the runoff generation potential of the catchment and by impounding part of the generated runoff into the cultivated fields. Often, afforestation activities are undertaken in the catchment by community organizations, which affect runoff generation. Trees such as eucalyptus which were preferred for plantation under afforestation programs were water-guzzlers. They can suck groundwater, apart from capturing part of the runoff and depleting soil moisture. Such changes are occurring everywhere in the rural landscape. But, there are two notable exceptions. (1) Places where the forests constitute the main catchment of tanks. (2) Areas where groundwater development is not feasible for agricultural intensification due to the presence of hard rock geology, and outcrops of underlying geological strata. Thus, in areas which have experienced significant changes in land use in the form of expansion of cultivation in the catchment, it will not be economically prudent to invest in tank rehabilitation.

The second type of change in the catchment land use is caused by the use of clayey soils in the catchment and tank bed for brick-making, etc. With booming construction activity in the state, there is mounting demand for bricks. The pressure on catchment land and tank bed for such uses is more in case of tanks which are located in the vicinity of towns. Such activities can also change the runoff or storage potential of the tank, depending on the place from where the soil is excavated. In such situations, the communities or the local Panchayats will not have much interest in reviving the tanks as the income earned from such activities is very large.

The third type of change is the interception of the drainage lines in the catchment. There are two different types of activities which cause this interception. (1) Construction of roads and buildings. (2) The indiscriminate construction of water-harvesting structures such as check dams. This is very rampant in Andhra Pradesh, like in many semiarid states of the country. The absence of any kind of regulations on water resources in the state has actually precipitated serious concerns. One of the reasons for this unprecedented increase in water-harvesting schemes is that the demand for water in agriculture has increased in the upper catchments of river basins with growing population pressure, and with depleting groundwater resources. In such areas, where such inflow reduction is clearly visible, it will not be economically prudent to invest in large-scale tank rehabilitation.

The degradation of tanks occurring as a result of changes in tank hydrology also seems to affect the success of the rehabilitation program. It was observed that in the case of Kurnool and Nizamabad districts, more money was spent for rehabilitation works of those tanks which are actually not performing well as compared with the good ones. In spite of this, the condition of poorly performing tanks did not improve.

In the final analysis, it appears that groundwater-intensive use in upper catchment or lower catchment will have the most remarkable impact on hydrology and performance of tanks. This is because pumping of water from wells, while reducing the groundwater outflows into streams, also leads to intensive use of land in the catchment for cultivation, which further leads to a reduction in runoff generation and in situ harvesting of a portion of that runoff for storage in soil profile. In a nutshell, tank management is increasingly becoming a hydro-institutional challenge.

7.8 CONCLUSIONS AND POLICY RECOMMENDATIONS

Groundwater intensive use affects the performance of tanks in terms of the area irrigated by them. In Andhra Pradesh, the impact of bore well irrigation on tank performance appears to be higher than that of open wells. Though there are factors other than area irrigated, which need to be considered for assessing the performance of tanks, for a temporal study this could form a useful indicator of tank performance. However, the adverse impact of intensive well irrigation on tank performance in terms of reduction in tank-irrigated area has not been uniform across districts.

Intensity of land use and density of wells in the catchment are major determinants of tank performance. Low density of wells and low land-use intensity in the catchment provide favorable conditions for sustaining the hydrological integrity of tanks and therefore their good performance. Increasing intensity of wells and expansion in area under cultivation in the catchment reduces the irrigation performance of tanks.

The moribund idea of taking up every tank for rehabilitation, on the premise that a local institution could be created to manage it, needs a critical review. There are many external factors responsible for deterioration of tanks, impacts of which cannot be reversed by these institutions. The protocol for tank rehabilitation should consider estimates of the actual yield potential of the tank catchment using rainfall–runoff models, followed by realistic estimation of the water demand in the command. In the case of cascade tanks, the upper catchment tanks should receive the highest priority in rehabilitation programs. Since such exercise for examining the feasibility of tank rehabilitation involve significant costs, some simple and quantitative criteria were identified to short-list tanks for detailed studies.

REFERENCES

Asian Development Bank (ADB), 2006. Rehabilitation and Management of Tanks in India: A Study of Select States. Asian Development Bank, Philippines.

Balasubramanian, R., Bromley, D.W., 2002. Mobilizing indigenous capacity: A portfolio approach to rehabilitating irrigation tanks in South India. University of Wisconsin Madison, Wisconsin. www.aae.wisc.edu/www/events/papers/balasubramanian.pdf.

Balasubramanian, R., Selvaraj, K.N., 2003. Poverty, Private Property and Common Pool Resource Management: The Case of Irrigation Tanks in South India. Working paper no. 2. South Asian Network for Development and Environmental Economics, Kathmandu, Nepal.

Dhawan, B.D., 1985. Output impact according to main irrigation sources: Empirical evidence from four selected states. Paper presented at INSA National Seminar on "Water Management—The Key to Development of Agriculture, 20–30 April 1986.

Easter, K.W., Palanisami, K., 1986. Tank Irrigation in India and Thailand: An Example of Common Property Resource Management. Staff paper P86-35. Department of Agricultural and Applied Economics, University of Minnesota, Minnesota.
Gunnell, Y., Krishnamurthy, A., 2003. Past and present status of runoff harvesting systems in dryland peninsular India: a critical review. Ambio 32 (4), 320–324.
Kajisa, K., Palanisami, K., Sakurai, T., 2004. Declines in the Collective Management of Tank Irrigation and Their Impact on Income Distribution and Poverty in Tamil Nadu, India. Foundation for Advanced Studies on International Development, Japan.
Karthikeyan, C., 2010. Competition and conflicts among multiple users of tank irrigation systems. In: Proceedings of the Fourteenth International Water Technology Conference. International Water Technology Association, Alexandria, Egypt, pp. 837–852.
Kumar, M.D., Ghosh, S., Patel, A., Singh, O.P., Ravindranath, R., 2006. Rainwater Harvesting in India: Some Critical Issues for Basin Planning and Research. Land Use and Water Resources Research 6 (1), 1–17.
Kumar, M.D., Patel, A., Ravindranath, R., Singh, O.P., 2008. Chasing a mirage: Water harvesting and artificial recharge in naturally water-scarce regions. Economic & Political Weekly 43 (35), 61–71.
Kumar, M.D., Sivamohan, M.V.K., Niranjan, V., Bassi, N., 2011. Groundwater Management in Andhra Pradesh: Time to Address Real Issues. Occasional Paper no. 4. Institute for Resource Analysis and Policy, Hyderabad.
Kumar, M.D., 2002. Making the Montage: Setting the Agenda and Priorities for Water Policy Research in India Report of the Annual Partners' Meet of IWMI-Tata Water Policy Research program. International Water management Institute, Anand, Gujarat.
Kumar, M.D., 2007. Groundwater Management in India: Physical, Institutional and Policy Alternatives. Sage Publications, New Delhi.
Mosse, D., 1999. Colonial and contemporary ideologies of 'community management': the case of tank irrigation development in south India. Modern Asian Studies 33 (2), 303–338.
Mosse, D., 2003. Rule of Water: Statecraft, Ecology and Collective Action in South India. Oxford University Press, USA.
Muralidharan, D., Athavale, R.N., 1998. Artificial Recharge in India, base paper prepared for Rajiv Gandhi National Drinking Water Mission, Ministry of Rural Areas and Development. National Geophysical Research Institute.
Narayanamoorthy, A., 2007. Tank irrigation in India: a time series analysis. Water Policy 9 (2), 193–216.
von Oppen, M., Rao, K.S., 1987. Tank Irrigation in Semi-arid Tropical India: Economic Evaluation and Alternatives for Improvement. Research Bulletin no. 10. International Crops Research Institute for the Semi-Arid Tropics, Patancheru, Andhra Pradesh.
Palanisami, K., Easter, K.W., 1991. Hydro-economic interaction between tank storage and groundwater recharge. Indian Journal of Agricultural Economics 46 (2), 174–179.
Palanisami, K., Meinzen-Dick, R., Giordano, M., 2010. Climate change and water supplies: Options for sustaining tank irrigation potential in India. Economic & Political Weekly 45 (26&27), 183–190.
Palanisami, K., 2006. Sustainable management of tank irrigation systems in India. Journal of Developments in Sustainable Agriculture 1 (1), 34–40.
Paranjape, S., Joy, K.J., Manasi, S., Latha, N., 2008. IWRM and Traditional Systems: Tanks in the Tungabhadra System. STRIVER policy brief no. 4. NIVA/Bioforsk, Norway.
PRADAN, 1996. Resource management of minor irrigation tanks and Panchyati Raj. Paper presented at the Seminar on Conservation and Development of Tank Irrigation for Livelihood Promotion, Madurai. July 12, 1996.
Prinz, D., 2002. The role of water harvesting in alleviating water scarcity in arid areas. In: Proceedings of the International Conference on Water Resources Management in Arid Regions, vol. III. Kuwait Institute of Science Research, Kuwait, pp. 107–122.

Rao, G.B., 1998. Harvesting water: irrigation tanks in Anantapur. Wasteland News 13 (3).
Sakthivadivel, R., Gomathinayagam, P., Shah, T., 2004. Rejuvenating irrigation tanks through local institutions. Economic and Political Weekly 39 (31), 3521–3526.
Sakurai, T., Palanisami, K., 2001. Tank irrigation management as a local common property: the case of Tamil Nadu, India. Agricultural Economics 25 (2&3), 273–283.
Sekar, I., Palanisami, K., 2000. Modernised rainfed tanks in South India. Productivity 41 (3), 444–448.
Shah, E., 2008. Telling otherwise a historical anthropology of tank irrigation technology in South India. Technology and Culture 49 (3), 652–674.
Shankari, U., 1991. Tanks: major problems in minor irrigation. Economic and Political Weekly 26 (39), A115–A124.
Sharma, A., 2003. Rethinking Tanks: Opportunities for Revitalizing Irrigation Tanks-empirical Evidence From Anantapur District. Andhra Pradesh, India. Working Paper no. 62. International Water Management Institute, Colombo, Sri Lanka.

CHAPTER 8

Reducing Vulnerability to Climate Variability: Forecasting Droughts in Vidarbha Region of Maharashtra, Western India

S. Deshpande
Groundwater Surveys and Development Agency, Pune, Maharashtra, India

N. Bassi
Institute for Resource Analysis and Policy, Delhi, India

M.D. Kumar
Institute for Resource Analysis and Policy, Hyderabad, India

Y. Kabir
UNICEF Field Office, Mumbai, Maharashtra, India

8.1 RATIONALE

In the Indian state of Maharashtra, groundwater is the main source of drinking water in rural areas. The tribal population in the state depends on hand pumps or dug wells to meet their domestic water needs and these sources dry up in peak summer. The situation is worse during droughts, when monsoon fails. When scarcity hits, the poor communities compromise their personal hygiene requirements to meet their productive water needs.

In Maharashtra, the Groundwater Surveys and Development Agency (GSDA) undertakes groundwater-level monitoring quarterly in 1531 watersheds to produce data on depth to water levels in wells and their seasonal fluctuations. Based on these data, the renewable groundwater resources available during the monsoon months in each of the watersheds is estimated. The existing monitoring system and the assessment methodology are not robust enough to capture all the real peaks and troughs of the water levels in wells, and the dynamic situation with respect to water availability in drinking water sources across the seasons. At best, they show the aggregate recharge occurring in watersheds and depth to water levels. In fact, the recharge assessment

Rural Water Systems for Multiple Uses and Livelihood Security
ISBN 978-0-12-804132-1
http://dx.doi.org/10.1016/B978-0-12-804132-1.00008-1

carried out periodically shows that very few watersheds (about 80) in Maharashtra experience groundwater overexploitation (GSDA and CGWB, 2013). However, in reality, hundreds of thousands of wells (agro wells and drinking water sources) have gone dry during the past 10–15 years in the state due to sharp seasonal drawdowns in water levels (Kumar et al., 2012).

High monsoon rainfall or large quantum of recharge during monsoon does not ensure year-round water availability in wells for domestic uses. In sum, there is no direct correlation between rainfall occurrence and groundwater availability and drought occurrence in a watershed. Therefore, drought forecasting cannot be made based on simplistic considerations of data on water level in wells and estimates of renewable groundwater resources in the watersheds, and instead should involve complex considerations.

There are many reasons for this. First, the groundwater withdrawal for agriculture from aquifers is very high in semiarid regions, as the demand for water in irrigated agriculture far exceeds the water demand for domestic uses in every village, and groundwater is the primary source of water to meet various needs in semiarid areas (Kumar et al., 2006). Second, the groundwater withdrawal for agriculture can change remarkably between a wet year and a dry year, as both the seasonal water demands and water availability can change dynamically between such typical rainfall years. Third, there is natural outflow of groundwater from shallow aquifers, and this is significant in hilly areas with hard rock formations and this is altered by withdrawals, due to which water-level trends in wells in such regions are quite dynamic and complex, and there are limits on utilizable recharge induced by the geohydrological setting of an area (Kumar and Singh, 2008; Kumar et al., 2012).

In order to integrate these complex considerations in assessing droughts and water scarcity, historical data for selected locations in a block falling in Vidarbha region of Maharashtra were correlated on the following: rainfall magnitude and pattern; runoff/streamflows, including lean season flows; depth to water levels in wells during different seasons; pre–post monsoon fluctuation in water levels in wells; the cropping pattern and cropped areas (in different seasons); and occurrence of droughts in terms of intensity and extent. The outcomes of these analyses can be applied to the data obtained from real-time monitoring of precipitation and water levels in observation wells to improve the predictability of "drought occurrence" and "emergence of drinking water shortage."

8.2 CURRENT APPROACHES TO DROUGHT FORECASTING

Drought is the most complex and frequently occurring climate-related disaster, causing significant damage both in the natural environment and in human lives, but is the least understood of all the natural hazards (Mishra and Desai, 2005; Rojas et al., 2011). One of the basic deficiencies in mitigating the effects of drought is the inability to forecast drought conditions reasonably well in advance (Mishra and Desai, 2005). However, drought forecasting is a critical component of drought hydrology, which plays a major role in risk management, drought preparedness, and mitigation. The different components of drought forecasting include: input variables (precipitation, streamflow, groundwater levels, soil moisture, crop yield, etc.); methodology, and the outputs obtained (Mishra and Singh, 2011).

Various methodologies have been developed for drought forecasting. These include: (1) regression analysis between two or more quantitative variables; (2) time series analysis using autoregressive integrated moving average (ARIMA) and seasonal autoregressive integrated moving average (SARIMA) models; (3) probability models such as Markov chain models; (4) artificial neural network models which are a class of flexible non-linear models that can adaptively discover patterns from the data; (5) hybrid models which are useful in extracting advantages of individual models for predicting droughts with better accuracy as well as for higher lead time in comparison to individual models; and (6) long-lead drought forecasting which uses climate indices as well as the periodic nature of hydro-meteorological variables (Mishra and Singh, 2011).

However, use of all these models has some limitations. For instance, regression analysis presents a difficulty in terms of understanding the underlying causal mechanisms and multicollinearity. Similarly, the use of artificial neural network involves lot of empirical information and plenty of computations (Mishra and Singh, 2011).

Nevertheless, no specific institutional arrangement exists for drought forecasting in India. At the national level, the Indian Metrological Department (IMD) prepares the weather reports which mainly forecast on the occurrence of rainfall over a region averaged over a particular time period. The government effort is mainly focused on assessing the impact of droughts and undertaking relief operations (mostly in the form of financial assistance) once it has happened.

8.2.1 Various Indices Used for Drought Assessment

Drought indices have been derived for assessing the effect of a drought and defining different drought parameters, such as intensity, duration, severity, and spatial extent (Mishra and Singh, 2010). Some of the important indices include: (1) Standardized precipitation index (SPI) which uses long-term precipitation data; (2) normalized difference vegetation index (NDVI) which analyses remote sensing measurements of vegetation; (3) Palmer drought severity index (PDSI) which is based on precipitation and temperature; (4) surface water supply index (SWSI) which includes snowpack, reservoir storage, streamflow, precipitation, and other measurements and is calculated separately for each river basin; and (5) standardized runoff index (SRI) which incorporates hydrologic processes that determine the seasonal loss in streamflow due to the influence of climate.

Most of these indices measure different drought-causative and drought–responsive parameters, and identify and classify drought accordingly. However, these parameters are not linearly correlated with each other and therefore it is important to investigate the consistency of results obtained by different drought indices (Awass, 2009).

8.2.2 Drought Assessment in Maharashtra

In Maharashtra, which is a drought-prone state, NDVI and normalized difference water index (NDWI) are mainly used for assessment of agricultural droughts. NDVI is derived as: NDVI = (NIR − Red)/(NIR + Red), where NIR (near infrared) and Red are the reflected radiation in visible and near-infrared channels. Various colors in the NDVI image (yellow, green, and red) indicate increasing vegetation vigor. The NDVI values for vegetation generally range from 0.1 to 0.6, the higher index values being associated with greater green leaf area and biomass (NCFC and NRSC, 2012).

NDWI is derived as: NDWI = (NIR − SWIR)/(NIR + SWIR), where NIR and SWIR (shortwave infrared) are the reflected energy in these two spectral bands. Higher values of NDWI signify more surface wetness (NCFC and NRSC, 2012).

NDVI is most widely used for operational drought assessment because of its simplicity in calculation, easiness in interpretation, and also its ability to partially compensate for the effects of atmosphere, illumination geometry, etc. (NCFC and NRSC, 2012). However, these indices are used only for post facto assessment of agricultural droughts and not for forecasting droughts (meteorological or agricultural), which is essential for the optimal operation of water supply systems and formulation of drought mitigation strategies.

8.3 PROJECT GOALS AND OBJECTIVES

As highlighted in Section 8.2, drought forecasting is still not that advanced in India. Thus, the main goal of the research project was to better management of groundwater resources locally for drinking water security through better prediction of drinking water scarcity for the most deprived sections of the society, the tribal population, based on real-time monitoring of daily rainfall and groundwater levels at high frequency. Towards this goal, the objective is to set up a web-based decision support tool (DST) which establishes a relationship between the rainfall characteristics (amount, number of rainy days, and onset and retreat of monsoon), condition of groundwater in terms of quantity, and drought occurrence and summer scarcity.

The innovation seeks to improve the predictability of quantity and quality of water in drinking water supply wells and droughts, using data on rainfall, groundwater levels, and cropping patterns. As a result, it will provide real-time information on the condition of groundwater in the villages.

8.4 FEATURES OF THE PROJECT AREA

The research project was undertaken in Jiwati block of Chandrapur district falling in the eastern part of Vidarbha region in Maharashtra. The block is located at about 55 km south of Chandrapur District headquarter. The geographical area of the block is about 505 square kilometers (sq. km). Forest area comprises about 254 sq. km and wasteland is 72 sq. km. Area under cultivation is 15,245 ha. The average rainfall in the block is about 1100 mm with high inter- and intra-annual variability. The temperature varies from a maximum of 45°C to a minimum of 30°C.

Chandrapur district is well drained by three tributaries of the Godavari river system. These are Wardha, Wainganga, and Penganga. These three rivers along with their tributaries rise in the upland within the district and drain the entire district. The selected block from the district is underlain by varied types of geological formations ranging from Proterozoic to Recent. Penganga limestone and shale are found along the northwest part of the block. These rocks are overlain by unclassified basaltic lava flows. The individual flow thickness varies from 25 to 30 m. The occurrence of groundwater is restricted in the weathering parts of basaltic rocks. As per the estimates of the Central Ground Water Board (CGWB) and GSDA, groundwater development in the block is within safe levels (CGWB, 2009). However, well failures are very common, with drying up of open wells during summer months.

There are a total of 83 villages in the selected block. As per the Census of India (2011), the total population is 61,820, out of which 29% belongs to schedule tribes (Census of India, 2011). Generally in the monsoon season, crops such as paddy, sorghum, maize, pigeon pea, green gram, black gram, soya bean, sesame, edible oil seeds, cotton, and vegetables are cultivated. In the winter season, wheat, sorghum, gram, vegetables, maize, pulses, and oil seeds are the major cultivated crops. However, area under winter crops, which require irrigation, is a small fraction of the total cropped area. Further, during summer, no crops are cultivated in the area due to the acute scarcity of water.

A high proportion of households in the block (almost 87%) primarily depends on groundwater-based sources (wells, bore wells, hand pumps, etc.) to meet their drinking water requirements (Census of India, 2011). However, as emerged during discussions with the village community members, groundwater availability for drinking purposes lasts only till the month of March. By April and May, water shortage is experienced for both drinking and irrigation purposes. Thus, a high proportion of rural households, which are dependent on groundwater-based sources, face acute water scarcity during 2 months of intense heat.

Furthermore, it emerged during the discussions that a general reduction in the amount of rainfall has been observed from 2005 to 2011. In 2009, a drought was declared in the block and many government schemes were started in order to address the issue of water shortage for drinking and irrigation purposes. However, none of these schemes were completed. Also, no fodder was provided to feed the affected livestock. However, the government does offer compensation ranging from INR 4000 to INR 7000[1] to each affected farmer in case standing crops are damaged due to non-monsoon rains.

8.5 PROJECT APPROACH

The proposed web-based DST envisages periodic collection of relevant data on predetermined indicators illustrating the status of groundwater resources. The data collection will be at the village level and will be communicated periodically by "SMS" technology to a central node for further tabulation and analysis for prediction of the groundwater situation, and probability of occurrence of droughts and summer scarcity. This will help inform the existing decision-makers for a more proactive response to ensure ground water safety and security.

The research project was also about developing methodologies for prediction of droughts, it is expected to benefit other areas of the district having similar agro-climatic and geomorphological settings, for: (1) predicting meteorological droughts; and (2) establishing a relationship between rainfall and water-level trends in wells (post monsoon and summer), and hydrological events (rainfall, recharge, and lean season outflows) and socioeconomic droughts.

[1] One USD is equal to 60 Indian National Rupees (INR).

8.6 SETTING UP OF DST

8.6.1 Methods

Developing a DST, which can predict the quantum of utilizable groundwater resources and thereby drinking water scarcity situation in a region, first requires a comprehensive understanding of the physical processes affecting water availability in aquifers. A major challenge is to know with some degree of reliability, how much water would be effectively available in the aquifer for exploitation toward the end of the hydrological year. Currently, the total groundwater pumping for irrigation and domestic uses and industry is compared against the net monsoon and non-monsoon recharge from precipitation and irrigation. But, the estimation of "net recharge" does not consider the natural outflows from the aquifer during the monsoon period (Kumar et al., 2012). Since the groundwater discharge (base flow) is likely to be significant in hilly areas, this has to be estimated.

To predict water-level fluctuations based on rainfall, arriving at a net monsoon recharge is important. But, the net recharge cannot be correlated with rainfall (due to the weak correlation that exists), but gross monsoon recharge, which is net recharge plus groundwater outflow during monsoon. Hence, groundwater outflows during this season need to be estimated. Given the likely stronger correlation between gross recharge and rainfall, one might be able to estimate the recharge coefficient (recharge/rainfall) for every year, based on the measured value of rainfall. The methodology needs to be worked out separately for estimation of the groundwater outflows for both the monsoon season and the lean season. The second would help estimate the effective groundwater availability for meeting various needs during the entire year. We will discuss the methodology for estimation of both the variables separately.

Simulation studies have established the relationship between rainfall and runoff for various rainfall magnitudes, with coefficients for surface runoff and base flows, separately, for some river basins of the Western Ghats (NIH, 1999). The topography of these basins is similar to that of river (sub)basins in Chandrapur district. Nevertheless, since the land cover and soil conditions (infiltration capacity and moisture regimes) in Chandrapur region are different from those of Western Ghats, the rainfall–runoff relationship developed for that region will not be applicable for Chandrapur. However, the coefficients for base flow (ie, what proportion of the runoff is contributed by the base flow), which is governed largely by the geomorphological conditions, can be of use, if estimated for each annual rainfall. However, this coefficient is not constant. It is found to vary with changing magnitude of rainfall in the western Ghat region.

While the runoff coefficient (K_r) is found to increase with rainfall magnitude, the base flow coefficient (K_u) is found to reduce, though the absolute value of the base flow increases. Hence, the base flow coefficient can keep changing with the annual rainfall. To estimate the runoff for a given rainfall occurrence in future, the rainfall–runoff relationship for the catchments in Chandrapur needs to be estimated. For this, regression was run between annual rainfall (in mm) and the observed streamflows, assuming that there is no major impoundment of the water before the gauging points in the rivers. Once the runoff is computed, the base flow component can be deduced. The base flow of lean season has to be subtracted from this to obtain the monsoon base flow.

The extent of natural groundwater outflow during the non-monsoon period (lean season base flow) can be directly deduced from the lean season flows in rivers draining the area, as there is likely to be no surface runoff during this period. The lean season flow, divided by the area of the catchment draining, can also be estimated and expressed as a coefficient of the annual rainfall.

The modeling tools were developed on the basis of time series data on rainfall, rainy days, groundwater levels, pre–post monsoon water-level fluctuations, cropping and irrigation intensity, and "drought occurrence" to predict the following on the basis of real-time monitoring of rainfall and groundwater-level fluctuations: occurrence of meteorological droughts; cropping and irrigation intensities; utilizable groundwater recharge; and extent of socioeconomic droughts in terms of number of villages affected by drinking water scarcity.

Time series data were collected on daily rainfall and rainy days for two locations for the period from 1960 to 2012, and on groundwater levels of GSDA's existing observation wells (a dug well data from 1985 to 2012 and a bore well data from 1999 to 2012) in the selected block. Streamflows data from 1990 to 2005 were also collected for a catchment (part of Wainganga sub-basin) of 1900 sq. km falling in the block. In addition, aggregate block level data on cropping pattern and cropped area in different seasons were collected for the period from 1992 to 2012 for whichever year the data are available. Data were also collected on drought occurrence and the number of villages affected by droughts. The following analyses were done for developing the modeling tool:

1. Analysis of probability of occurrence of rainfalls of different magnitudes based on the historical rainfall data; and drought frequency analysis of using standard precipitation index (SPI).

2. Analysis of relationship between number of rainy days and annual rainfall.
3. Analysis of relationship between the onset of monsoon and occurrence of meteorological drought.
4. Regression analysis involving total annual rainfall (independent variable) and pre–post monsoon groundwater-level fluctuations (as dependent variable): to analyze the impact of rainfall magnitude on the quantum of monsoon recharge in the watersheds.
5. Regression analysis of rainfall and runoff for developing the rainfall–runoff model.
6. Methodology for estimating total rainfall infiltration from annual rainfall values, which is based on the estimates of monsoon recharge and groundwater discharge during monsoon.
7. Regression analysis involving data on total annual rainfall and "drought occurrence": to examine whether any relationship exists between meteorological drought[2] and socioeconomic drought[3].
8. Regression analysis involving total annual rainfall, and cropping and irrigation intensities to examine how rainfall variations influence cropping decisions.
9. Regression analysis involving total annual rainfall and peak summer water levels in wells.

These analyses help to understand the inter-relationship of various physical processes (precipitation, occurrence of meteorological drought, water-level trends, and groundwater recharge) and relationship between physical processes and socioeconomic processes (changes in cropped area and cropping pattern shift) in determining the final impacts in terms of droughts and drinking water shortage. These models together constitute the DST.

8.6.2 Modeling Results

The major statistical models developed on the basis of the analyses discussed in Section 8.6.1 are discussed below.

8.6.2.1 Relationship Between Annual Rainfall and Rainy Days

The analysis showed a strong linear relationship between the two. As the rainfall increased, the number of rainy days also increased proportionately (Fig. 8.1). The difference between the mean number of rainy days and the

[2] Meteorological drought refers to departure of annual precipitation from the normal values. Depending on the magnitude of rainfall as a fraction of the normal values, it would be classified as moderate drought and severe drought.

[3] Socioeconomic drought refers to crop failures, fodder shortage, and drinking water scarcity during monsoon months, which can be the result of monsoon failure.

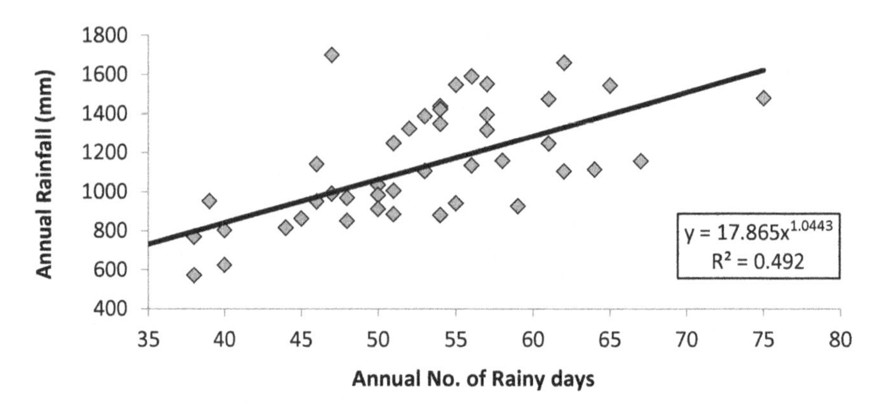

Figure 8.1 Relationship between rainfall and rainy days (1960–2012) for Jiwati block.
Source: Authors' own analysis.

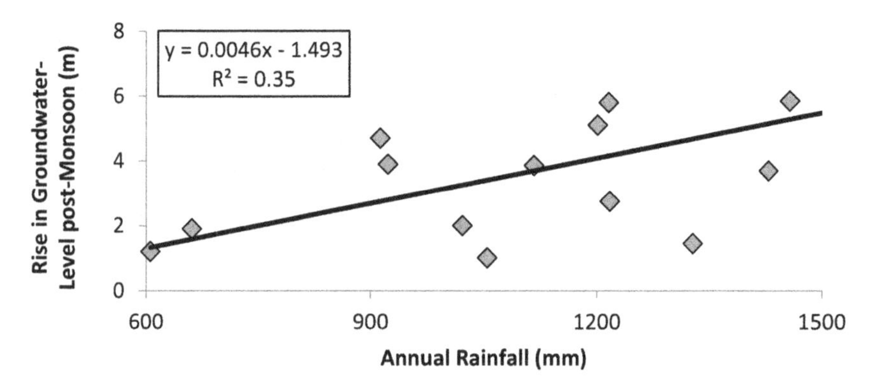

Figure 8.2 Relationship between water level (bore well) and rainfall in Jiwati block.
Source: Authors' own analysis.

amount of rainfall amongst the above normal and below normal rainfall years is 10 (rainy days) and 505 mm, respectively. This suggests that even a non-occurrence of a single-day rainfall significantly affects the amount of total annual precipitation. This is consistent with the pattern emerging vis-à-vis the relationship between rainfall–rainy days for the country as a whole.

8.6.2.2 Relationship Between Rainfall and Water-Level Fluctuations in Open Wells and Bore Wells

The relationship between annual rainfall and pre- and post-monsoon groundwater-level fluctuation (Fig. 8.2) was found to be stronger for water levels in bore wells, while the water-level fluctuations in open wells were less sensitive to the rainfall, with a weak relationship between the

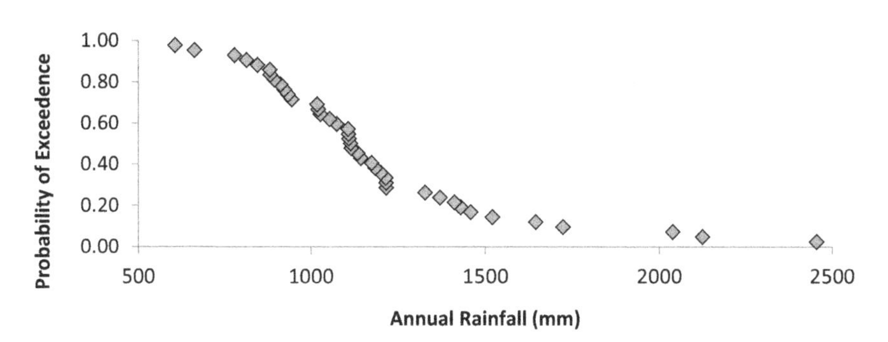

Figure 8.3 Probability of exceedance of a rainfall event in Jiwati block. *Source: Authors' own analysis.*

rainfall and water-level fluctuation in wells. The reason for the latter could be due to the fact that groundwater outflows from shallow aquifers (which the open wells tap) during the monsoon season is very significant. The summer water levels almost remained steady, irrespective of the rainfall magnitude. This is probably due to the reason that, in good years, the natural discharge increased with increase in magnitude of rainfall, and the limited residual storage in the aquifer gets fully depleted in both good and bad years.

8.6.2.3 Assessing the Probability of Occurrence of Rainfall of Different Magnitudes

The probability curves were estimated for annual rainfall to find out the minimum quantum of rainfall that can occur at different probabilities, like 50%, 60%, and 75%. This used rainfall data for 53 years from 1960 to 2012 (Fig. 8.3). The analysis showed that the region can experience a rainfall as high as 2450 mm once in nearly 42 years.

8.6.2.4 Estimation of Standard Precipitation Index (SPI) and Drought Frequency Analysis

The intensity of occurrence of droughts was also estimated in terms of standard precipitation index (SPI), by examining the extent to which the precipitation in a particular year departed from the mean values in terms of number of standard deviations. Subsequently, the probability of occurrence of droughts of different intensities was also estimated, by estimating the probability curve for SPI. The probability of occurrence of a severe meteorological drought is only 6% and the probability of occurrence of moderate drought is around 15% (Fig. 8.4).

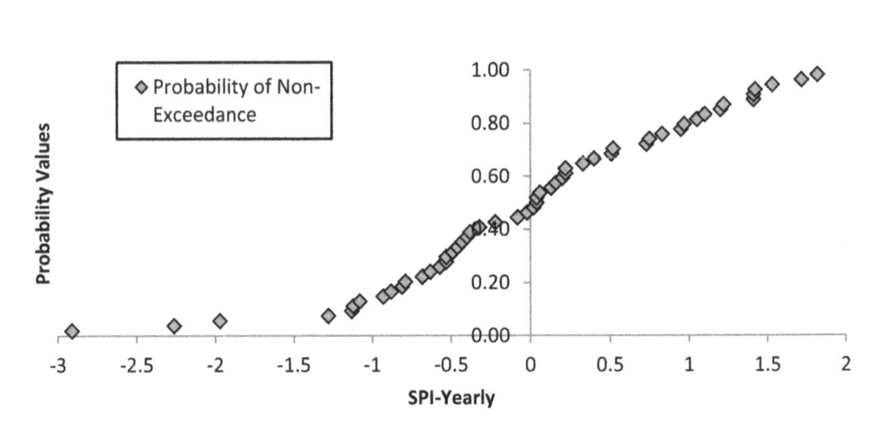

Figure 8.4 Probability of occurrence of droughts of different intensities in Jiwati block. *Source: Authors' own analysis.*

8.6.2.5 Identifying the Factors Determining Water-Level Fluctuation in Wells During the Monsoon and the Factors Influencing the Net Groundwater Balance

Analysis of time series data on groundwater-level fluctuations during monsoon and water-level trends after the monsoon has shown base flow during monsoon to be an important factor, determining the water-level fluctuation in the aquifer due to rainfall. Furthermore, groundwater discharge during the lean season was found to be an important factor deciding the water-level situation in wells during summer.

8.6.2.6 Estimation of Total Rainfall Infiltration From Rainfall

Water-level fluctuation in wells during monsoon and the net monsoon recharge cannot be predicted using rainfall data owing to the tenuous relationship between the two in hilly, hard rock areas. Therefore, indirect methods have to be employed for estimating utilizable recharge. A hydrological parameter, which fully depends on rainfall magnitude, is total infiltration, and therefore can be predicted using rainfall figures. From this utilizable recharge can be estimated, as total infiltration is the sum of the "net recharge during monsoon" and the total groundwater discharge (base flow). Here an important unknown for most catchments is "groundwater outflow."

For catchments with similar topographical and geohydrological characteristics, the relationship between base flow and the runoff is constant for a given rainfall. Hence, in order to estimate the groundwater outflow during monsoon, a validated model for rainfall–base flow coefficient (ratio of base flow and runoff) for a similar catchment in the Western Ghats was

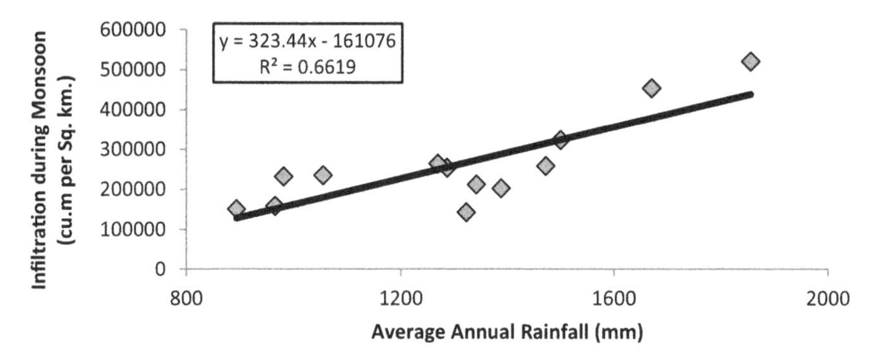

Figure 8.5 Contribution of annual rainfall to gross monsoon infiltration in Jiwati block. *Source: Authors' own analysis.*

used to estimate the total base flow in the catchment under study. From the available data on observed streamflows for a particular catchment, the total groundwater outflow corresponding to each rainfall was estimated using these base flow coefficients obtained from interpolation. The estimates of base flow during the monsoon were obtained by subtracting the lean season streamflows, which is nothing but the base flow during the non-monsoon period. From the estimated values of total infiltration and annual rainfall, a relationship was obtained and the model was found to be robust. The total infiltration from rainfall increased with the magnitude of rainfall (Fig. 8.5).

8.6.2.7 Estimation of Utilizable Recharge From Measured Values of Annual Rainfall

In order to predict the utilizable recharge from future rainfall events, first a rainfall–runoff model was developed using average rainfall and streamflow data for Wainganga sub-basin of the Godavari which falls in the area under study. The regression equation had an R^2 value of 0.67 (Fig. 8.6). Using this "rainfall–runoff model," the annual runoff for known values of future rainfalls could be estimated. From the estimated values of runoff, the total groundwater discharge/outflow could be estimated using the "base flow coefficient" determined on the basis of annual rainfall. The net utilizable recharge can then be estimated by subtracting the total groundwater outflow from the "total infiltration," the latter being deduced from the regression model linking "annual rainfall with total infiltration."

Nevertheless, the current "rainfall–runoff model" could be refined using more accurate values of average annual catchment rainfall, which could

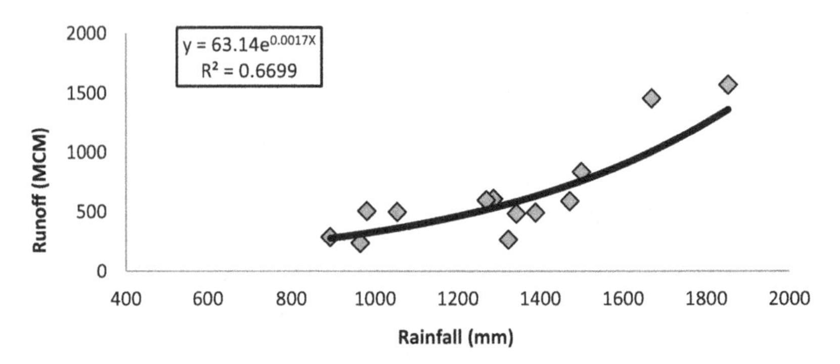

Figure 8.6 Rainfall–runoff relationship for a catchment (part of Wainganga sub-basin) falling in Jiwati block. *Source: Authors' own analysis.*

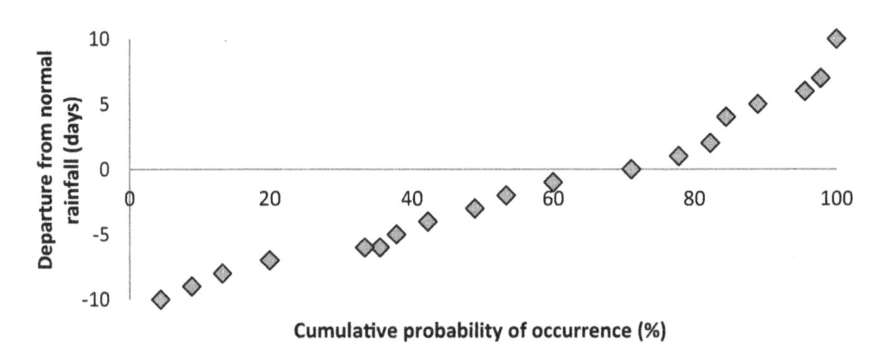

Figure 8.7 Probability of rainfall advancement in Jiwati block. *Source: Authors' own analysis.*

capture the potential spatial variations in rainfall across the catchment and longer duration time series data on observed flows.

8.6.2.8 Predicting the Occurrence of Meteorological Droughts

The normal date of arrival of monsoon in Jiwati block of Chandrapur is June 11. Analysis showed that there is greater probability of early arrival of monsoon in the area than late arrival. Fig. 8.7 shows the cumulative probability of rainfall advancement. The probability of arrival of monsoon earlier than the normal date is as high as 71%. Furthermore, early arrival of monsoon in the area increases the probability of occurrence of droughts. Empirical analysis showed that there is a greater incidence of meteorological droughts, assessed in terms of SPI, during years of early arrival of monsoon than during years of late arrival. This runs contrary to the general belief that delayed arrival of monsoon leads to rainfall deficit.

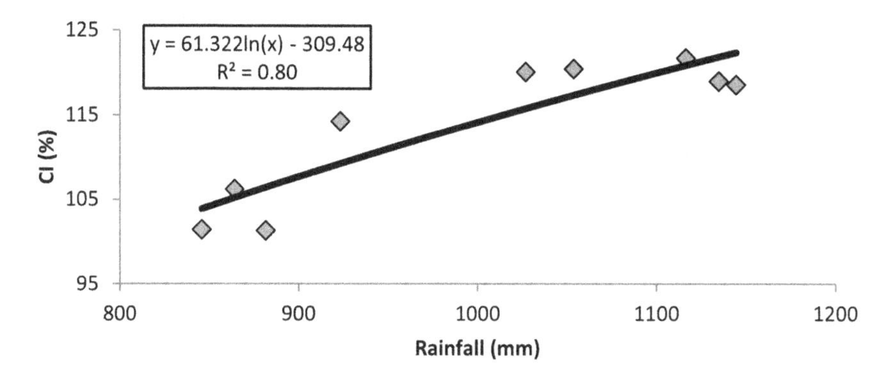

Figure 8.8 Impact of rainfall on cropping intensity in Jiwati block. *Source: Authors' own analysis.*

8.6.2.9 Predicting the Cropping and Irrigation Intensity Trends From Measured Values of Rainfall

In poorly endowed regions which lack adequate irrigation infrastructure, especially those which involve transfer of water from water-rich regions, crop production will be heavily dependent on the monsoon performance. Chandrapur is one such region. Analysis was performed to examine how annual rainfall impacted on crop production in terms of cropping intensity and irrigation intensity, using historical data. The analysis showed a strong direct relationship, with the increase in rainfall magnitude increasing the cropping intensity and irrigation intensity. The models are robust, with rainfall explaining variation in cropping intensity in the order of 80% (Fig. 8.8), and irrigation intensity in the order of 78%. Nevertheless, the area cropped during the winter season (which is irrigated) remains a small fraction of the net sown area even in good rainfall years due to the reason that the utilizable groundwater recharge is very limited. The next level of analysis should involve the actual agricultural outputs in value terms, as the biggest impact of rainfall variations in such regions is likely to be on the monsoon season yield and production.

8.6.2.10 Predicting Summer Water Levels and Drinking Water Scarcity Based on Known Trends in Rainfall

The analysis of rainfall and depth to water levels in summer showed a linear, but inverse, relationship. The higher the rainfall, the lower was the depth of water levels in wells during summer. However, the reduction for a 100-mm increase in rainfall is very insignificant, suggestive of the phenomenon of "dynamic equilibrium" in groundwater conditions.

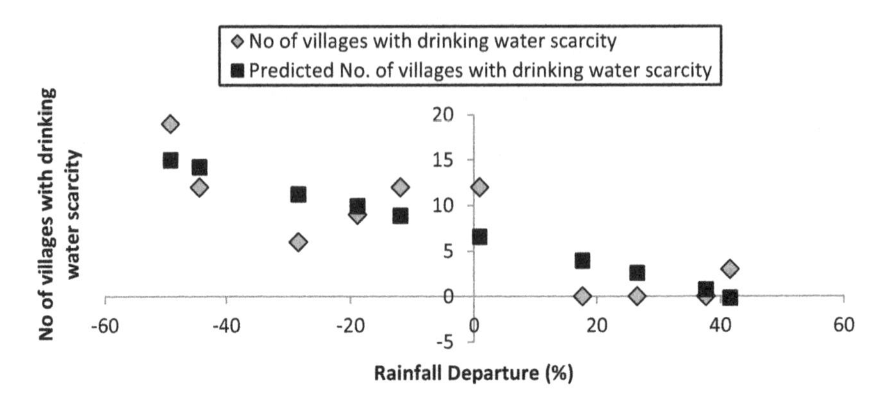

Figure 8.9 Rainfall departure and extent of drinking water scarcity in Jiwati block. *Source: Authors' own analysis.*

As expected, the advancement/delay in monsoon also had some impact on the drinking water situation—expressed in terms of number of villages affected by summer water scarcity. The greater the advancement in arrival of monsoon (in number of days), the higher was the extent of summer water scarcity. It is also important to reckon with the fact here that advancement of monsoon also coincided with a higher incidence of meteorological droughts, than delay in monsoon. However, the data on drinking water scarcity is available for only a few years, and therefore, the model has poor "predictability."

At the same time, the relationship between the departure of rainfall from normal values (% change in rainfall from normal values) and the incidence of drinking water scarcity was much stronger and "inverse linear." In years of high monsoon deficit, larger numbers of villages were reported to be affected by drinking water shortage during summer, and in years of good monsoon, no villages were reported to be affected by water shortage (Fig. 8.9).

8.6.2.11 Estimating Groundwater Discharge During Lean Season

Since the real-time groundwater monitoring gives figures of water-level fluctuations, these figures can be used to estimate the net monsoon recharge. The gross infiltration during monsoon can be obtained using the model we have established. The difference between the two will give the groundwater outflow during monsoon (gross infiltration minus net recharge). The total groundwater outflow can be estimated from the annual runoff estimates using the "base flow coefficient," determined on the basis of rainfall. The difference between "total base flow/total groundwater outflow" and the outflow or discharge during the monsoon would yield estimates of "lean season flow" in the streams. The lean season flow can also be measured using stream gauging at appropriate locations in the catchment.

8.7 CONCLUSIONS AND AREAS FOR FUTURE WORK

The existing groundwater monitoring system is not robust enough to capture all the spatial variation in fluctuation in water levels in wells due to poor monitoring networks, whereas, the official estimates of groundwater recharge do not provide the dynamic situation with respect to water availability in wells across the seasons. At best, they show the aggregate recharge occurring in watersheds. In fact, there is no direct correlation between rainfall occurrence and groundwater availability and drought occurrence in a watershed. Therefore, prediction of droughts and drinking water scarcity cannot be made based on simplistic considerations of data on water level in wells and estimates of renewable groundwater availability in the watersheds and instead should involve complex considerations.

In order to integrate these complex considerations in assessing droughts and water scarcity, we have tried to understand the system behavior by looking for critical relationships that exist between various physical processes affecting groundwater availability in the local aquifers and between these physical processes and their socioeconomic outcomes. For this, we have run several regression models and other statistical analyses using the historical data for selected locations in a block of Chandrapur district on the following: rainfall magnitude and pattern; date of arrival of monsoon; pre–post monsoon water-level fluctuation in wells and depth to water levels in different seasons; annual streamflows and lean season flows; the cropping and irrigation intensity; and extent of occurrence of droughts in terms of number of villages affected.

Through these analyses, we have established several critical relationships, which are important for drought predictions. They are relationships between: rainfall and rainy days; rainfall and groundwater-level fluctuations during monsoon; rainfall and streamflows; rainfall and cropping and irrigation intensity; date of onset of monsoon and the probability of occurrence of droughts; and rainfall and extent of drinking water scarcity. We have also estimated the probability of occurrence of droughts of different intensities and probability of early and delayed arrival of monsoon in the study area. More importantly, we have also found the reasons why monsoon water-level fluctuations cannot be predicted merely on the basis of rainfall, and identified the key hydrological variables influencing the groundwater behavior in response to rainfall, and worked out practical methodology for estimating these variables. In the process, we have refined the current methodology for assessment of utilizable groundwater recharge from rainfall, which incorporates the complex processes affecting the available groundwater other than "pre–post monsoon water-level fluctuations." The host of models would work as a decision support tool for drought predictions and management of drinking water crises during summer.

The models can predict the following: the probability of occurrence of meteorological droughts in a particular year, based on date of arrival of monsoon. Furthermore, based on data of rainfall and water-level fluctuations in wells during monsoon, it can estimate the following: the quantity of groundwater in the aquifer which is utilizable from annual precipitation; runoff from the catchment; and, base flows, including lean season flows. It can also predict: the cropping and irrigation intensities; intensity of droughts in terms of number of villages likely to be affected by droughts; and summer water levels in wells. Hence, it can be an effective decision support tool for planning drought relief work, soon after the monsoon, as the data on actual rainfall in the area and water-level fluctuations arrive.

There is scope for refining these models for enhancing their performance in terms of the accuracy of predictions. For instance, the rainfall–runoff model could be refined using more accurate values of average annual catchment rainfall, which could capture the potential spatial variations in rainfall across the catchment and longer duration time series data on observed flows. Also, the data on streamflows should be of longer duration than what is available at present, and also modified to take into account all significant water diversions upstream so as to reflect the actual runoff (virgin flows). The data used currently for predicting changes in cropping intensity and irrigation intensity do not distinguish between rain-fed area and irrigated area, and therefore the area cropped during winter was assumed as irrigated area. Therefore, the data on cropping pattern, cropped area, and irrigated area in different seasons need to be obtained for longer durations, which in turn could be used to predict the changes in cropping decisions by the farmers in response to rainfall changes more accurately. Finally, the experimental data on rainfall–runoff–base flows need to be generated for pilot catchments of the region for a sufficiently large number of years at different spatial scales through hydrological monitoring so as to improve the accuracy of utilizable groundwater recharge assessments. Yet the most remarkable outcome of the study is development of a practical methodology for realistic assessment of utilizable groundwater recharge from rainfall in areas with undulating topography (IRAP, GSDA & UNICEF, 2015).

REFERENCES

Awass, A.A., 2009. Hydrological Drought Analysis-Occurrence, Severity, Risks: The Case of Wabi Shebele River Basin, Ethiopia. Dissertation Report. University of Siegen, Germany.

Census of India, 2011. Households by Main Source of Drinking Water and Location, Maharashtra. Ministry of Home Affairs, Government of India, New Delhi.

Central Ground Water Board (CGWB), 2009. Ground Water Information Chandrapur District Maharashtra. CGWB Central Regional Office, Nagpur, Maharashtra.

Groundwater Surveys and Development Agency (GSDA) & Central Ground Water Board (CGWB), 2013. Report of the Dynamic Groundwater Resources of Maharashtra (2011–2012). Groundwater Surveys and Development Agency and Central Ground Water Board, Pune and New Delhi.

Institute for Resource Analysis and Policy (IRAP), Groundwater Survey and Development Agency (GSDA) and UNICEF, 2015. Using Technology to Ensure Ground Water Safety and Security in a Tribal Block of Chandrapur, Maharashtra. Institute for Resource Analysis and Policy, Hyderabad.

Kumar, M.D., Singh, O.P., 2008. How serious are groundwater over-exploitation problems in India? A fresh investigation into an old issue. In: Kumar, M.D. (Ed.), Managing Water in the Face of Growing Scarcity, Inequity and Declining Returns: Exploring Fresh Approaches. Proceedings of the 7th Annual Partners' meet of IWM-Tata water policy research program, ICRISAT, Hyderabad, pp. 298–317.

Kumar, M.D., Ghosh, S., Patel, A., Singh, O.P., Ravindranath, R., 2006. Rainwater harvesting in India: some critical issues for basin planning and research. Land Use and Water Resources Research 6 (1), 1–17.

Kumar, M.D., Sivamohan, M.V.K., Narayanamoorthy, A., 2012. The food security challenge of the food-land-water nexus in India. Food Security 4 (4), 539–556.

Mishra, A.K., Desai, V.R., 2005. Drought forecasting using stochastic models. Stochastic Environmental Research and Risk Assessment 19 (5), 326–339.

Mishra, A.K., Singh, V.P., 2010. A review of drought concepts. Journal of Hydrology 391 (1), 202–216.

Mishra, A.K., Singh, V.P., 2011. Drought modeling–A review. Journal of Hydrology 403 (1), 157–175.

National Crop Forecast Centre (NCFC) and National Remote Sensing Centre (NRSC), 2012. Agricultural Drought Assessment Report. *Mahalanobis* National Crop Forecast Centre, Department of Agriculture & Cooperation and National Remote Sensing Centre, ISRO, Department of Space, New Delhi and Hyderabad.

National Institute of Hydrology (NIH), 1999. Rainfall-Runoff Modelling of Western Ghat Region of Karnataka. National Institute of Hydrology, Roorkee, India.

Rojas, O., Vrieling, A., Rembold, F., 2011. Assessing drought probability for agricultural areas in Africa with coarse resolution remote sensing imagery. Remote Sensing of Environment 115 (2), 343–352.

CHAPTER 9

Sustainable Access to Treated Drinking Water in Rural India

S. Bandyopadhyay
Associate Professor, School of Ecology and Environment Studies, Nalanda University, Rajgir, Bihar, India

9.1 INTRODUCTION

One of the Millennium Development Goals (MDGs) set a target to halve the proportion of the population without sustainable access to safe drinking water (and basic sanitation). Indeed, the target was met 5 years ahead of schedule. Globally, about 2.6 billion people have gained access to improved drinking water since 1990, of which 1.9 billion have gained access to piped drinking water in their dwellings. Nearly 58% of the world enjoys this higher level of service today. However, it is estimated that 663 million people worldwide still use unimproved drinking water sources, including unprotected wells, springs, and surface water bodies. Nearly half of all people using unimproved sources live in sub-Saharan Africa, while one-fifth live in South Asia (United Nations, 2015).

India, too, has met its target for access to clean drinking water, ensuring coverage for over 90% (UN India, 2015). Efforts will continue, and rightly so, to ensure universal, sustainable access to clean drinking water for rural India. Indeed the Sustainable Development Goals (SDGs), a follow-up of the MDGs, target universal availability and sustainable management of water.

However, sustainable health and other development benefits from universal access to water will only be realized when the complex issue of water safety and purity is appreciated and addressed in a sustainable manner. While there might be justifiable reasons to rejoice the achievement of certain targets, new challenges are emerging, urging a rethink of the way potable drinking water in rural India needs to be approached.

This chapter highlights the importance of quality of drinking water, as distinct from mere access to water, and opportunities for ensuring its sustainable access in rural India.

Rural Water Systems for Multiple Uses and Livelihood Security
ISBN 978-0-12-804132-1
http://dx.doi.org/10.1016/B978-0-12-804132-1.00009-3

9.2 INDIA WATER LANDSCAPE: CHALLENGES AND EMERGING TRENDS

Water resource availability in India is primarily governed by the seasonal rains associated with monsoons, topography, climate and geology, and is therefore subject to significant geographical variations in terms of rainfall patterns, surface runoff and storage, and groundwater availability. Moreover, there are important cultural, social, political, and (increasingly) economic drivers that influence patterns of access and use of water for agriculture, domestic purpose, livestock consumption, industry, and the environment.

Rural drinking water has traditionally been viewed as a social issue. The abiding image of women's drudgery has continued to define access as the key problem for potable water in rural India, as in other parts of the developing world. Loss of opportunity for education and productive work along with gender inequality and empowerment remained natural corollaries of this drudgery.

Access, therefore, has been defined as the key goal in rural drinking water over the past several decades. Indeed, as the access gap closes, it is the poorest who remain deprived, reinforcing commitments to social programs focused on universal access. While the economic viability gap increases as rural drinking water projects shift toward remote locations, often sparsely populated by the poor, there are a couple of other factors that add further complexity.

First, environmental degradation and increasing abstraction for competing use are sharply reducing freshwater availability for drinking and domestic uses. In addition, even this reduced availability is becoming uncertain, given the changes in climate systems as well as lack of clarity on ownership and use over water resources.

Second, and more importantly, groundwater sources are increasingly revealing a wide range of chemical contaminants in large parts of the country, posing a new set of challenges for public health and productive capacity of poor rural consumers (Kumar and Shah, 2004).

The MDG report card itself provides the first set of evidence of this complex interplay between poverty, health, and water use. While under-5 mortality rates are almost twice as high for children from among the poorest 20% of households, the likelihood of stunting is more than twice that compared with children from the wealthiest 20% of households. While drinking water might be only one of the many factors (another important factor being availability of skilled birth attendants) contributing to child mortality, it is definitely one of the major factors contributing to impaired growth and subsequent quality of life. Children in the poorest households are four times as likely to be out of school as compared to the wealthiest households.

The poor in rural India are, thus, faced with a Hobson's choice in terms of the quality of their drinking water. Institutional supplies are focused on access, while individual strategies typically risk contamination from consumption of surface water, water from an open dug well or from shallow tube wells, albeit to varying degrees.

9.3 GROWING NEED FOR TREATED DRINKING WATER IN RURAL AREAS

The decadal Census of India 2011 data reveal an interesting statistic—nearly two-thirds of households in India now have access to telephone services, while less than half have access to a toilet facility and less than a third use treated water for drinking. Almost 30% of urban households and 90% of rural households depend on untreated surface or groundwater (Srikanth, 2009).

The need for safe drinking water assumes greater significance when considered along with a hugely deficient public infrastructure for water and sanitation, rapid and unplanned growth of habitations, and poor hygiene practices in general. With less than 22% of the rural population having access to a latrine,[1] sanitation is extremely poor. Handwashing with soap is also very low, and continues to be a problem in India.

In India, water- and sanitation-related diseases account for between 70% and 80% of the burden of disease,[2] while the World Bank estimates 21% of communicable diseases to be related to unsafe water. The World Health Organization (Prüss-Üstün et al., 2008) estimates that 37.7 million Indians are affected by waterborne diseases annually—70–80% of the total disease burden in India, with 780,000 deaths attributable to contaminated water, and over 400,000 attributed to diarrhea alone. The disability-adjusted life years (DALYs) attributable to water, sanitation, and hygiene (WASH) was 28.2 million, with diarrhea accounting for over 13.6 million. The resulting economic burden was estimated at US$ 600 million per annum, including a loss of 73 million working days.

Infections are caused by a range of pathogens including viruses, bacteria, protozoans, and worms that find their way into drinking water that is

[1] Source: http://www.ddws.gov.in/sites/upload_files/ddws/files/pdf/BrochureonNurturingRural-India.pdf.

[2] Source: Tenth Five Year Plan, 2002–07. http://www.whoindia.org/LinkFiles/MDG_Chapter-05.pdf.

contaminated by sewage. Common diseases include hepatitis A, typhoid and paratyphoid fevers, giardiasis, amebiasis, cholera, and rotavirus diarrhea, etc. In India, the most common manifestation of waterborne infections is acute diarrheal disease (ADD), a basket of symptoms that include diarrhea, enteric fever, and cholera. In 2009 there were around 12 million reported cases of ADD (Health Status Indicators, 2010).

Cholera, originally endemic to the Ganga basin, is an extremely virulent disease, killing both children and adults within hours and easily transmitting and infecting others through the fecal–oral route. The disease has killed millions of people in South Asia since the 19th century and has even spread to Africa and the Americas. Recent advances in medical sciences, such as antibiotics and oral rehydration therapy, have reduced mortality from this killer epidemic significantly. Nevertheless, it is estimated that globally there are 1.4–4.3 million cases of cholera every year, leading to 28,000–142,000 deaths (Ali et al., 2012).

Amongst the other infective diseases linked to contaminated food and water, the most spectacular ones include dracunculiasis (guinea worm disease), schistosomiasis, and poliomyelitis, prevalent in countries of South Asia and sub-Saharan Africa. Other common helminth infections through food and water include roundworms, flukes, and tapeworms. While there is a close association between hygiene, sanitation, and water treatment at the source or point of use, it has been shown that water quality improvements have strong effectiveness leading to a 42% relative reduction in diarrhea on average in developing countries (Waddington et al., 2009).

Nearly 80% of India's rural drinking water today comes from underground sources, and hence it is considered to be generally safe in terms of pathogenic contamination. However, data submitted in Parliament by the water resources ministry[3] shows high salinity in 158 out of the 639 districts. In 267 districts, groundwater contains excess fluoride; nitrates are beyond permissible levels in 385; arsenic is present in 53; and there are high levels of iron in 270. In addition, aquifers in 63 districts contain heavy metals like lead, chromium, and cadmium, the presence of which in any concentration poses a danger.

The IMIS report[4] by NRDWP shows that about 15,565 habitations in 18 states have fluoride contamination in their drinking water, with five states having more than 1000 habitations with fluoride contamination. As shown in Fig. 9.1, iron is the most common contaminant, present in 27 states with

[3] Source: http://articles.timesofindia.indiatimes.com/2012-05-02/pollution/31537647_1_groundwater-nitrates-aquifers.

[4] Source: IMIS Report, NRDWP. http://indiawater.gov.in/imisreports/Reports/Physical/rpt_RWS_NoOfQualityAffHabitations_S.aspx?Rep=0&RP=Y&APP=IMIS (accessed 03.01.14.).

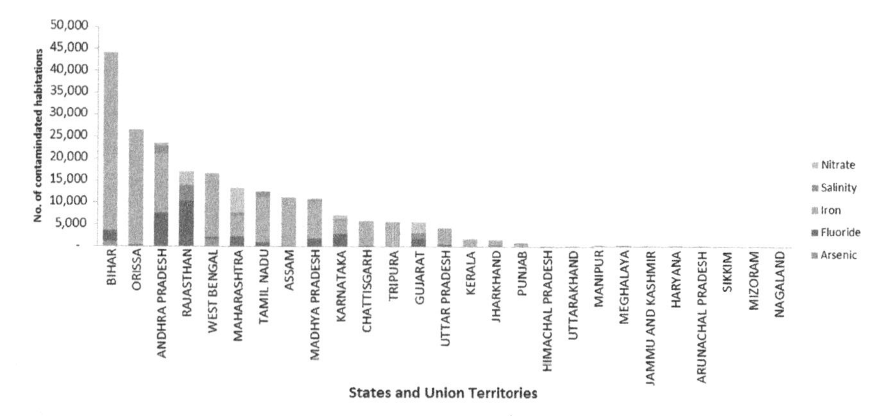

Figure 9.1 Number of contaminated habitations for five major contaminants in 2013–14. *Source: IMIS Report by NRDWP.*

Table 9.1 Indian states with chemically contaminated water sources

Contamination	States with one or more affected districts
Salinity	J&K, Himachal Pradesh, Uttarakhand, Northern Uttar Pradesh, Kerala, Sikkim, Chhattisgarh, Orissa, Western Ghats of Maharashtra and Karnataka and North-Eastern states
Fluoride	Andhra Pradesh, Gujarat, Karnataka, Madhya Pradesh, Rajasthan, Chhattisgarh, Haryana, Orissa, Punjab, Haryana, Uttar Pradesh West Bengal, Bihar, Delhi, Jharkhand, Maharashtra, and Assam
Iron	Andhra Pradesh, Assam, Bihar, Chhattisgarh, Goa, Gujarat, Haryana, Jharkhand, Karnataka, Kerala, Madhya Pradesh, Maharashtra, Manipur, Meghalaya, Orissa, Punjab, Rajasthan, Tamil Nadu, Tripura, Uttar Pradesh, West Bengal, and Andaman and Nicobar
Arsenic	Assam, Bihar, Chhattisgarh, Uttar Pradesh, and West Bengal

around 150,000 contaminated habitations. Nitrate is found in 17 states with about 14,000 habitations affected, salinity[5] affects about 9000 habitations in 14 states and arsenic contamination affects 4000 habitations in 13 states.

These chemicals have appeared in the water sources either due to too much water being drawn from increasing depth, or due to pollution from industrial and/or human waste. It is very often also caused by mineralization of groundwater. Some key contaminants include iron, fluoride, salinity, nitrate, and arsenic. Table 9.1 provides an overview of the states affected by each of these major contaminants (CGWB, 2010).

[5] The IMIS report does not give any limit of concentration values for salinity as Total Dissolved Solids.

Table 9.2 Major contaminants with the desirable and permissible limit as per BIS for drinking water IS 10500:2012 (second revision), health impacts, and sources[7]

Parameter	Requirement (desirable limit in mg/L)	Permissible limit (mg/L)	Health impacts	Sources
Arsenic	0.01	0.05	Weight loss; depression; lack of energy; skin and nervous system toxicity	Geological settings, previously used in pesticides (orchards) Improper waste disposal or product storage of glass or Electronics
Fluoride	1.0	1.5	Brownish discoloration of teeth, bone damage, fluorosis	Geological settings, industrial waste
Iron	0.3	No relaxation	Brackish color, rusty sediment, bitter or metallic taste, brown-green stains, iron bacteria, discolored beverages	Geological settings, leaching of cast iron pipes in water distribution systems
Nitrate	45	No relaxation	Methemoglobinemia or blue baby disease in infants	Natural deposits, decaying plant deposits, livestock waste, septic systems, manure lagoons, fertilizers, and household wastewater
Total dissolved solids	500	2000	Objectionable taste to water May affect osmotic flow and movement of fluids; at very high levels, kidney stones	Natural, reliance on deeper groundwater, dissolved minerals, nature of soil, and landfills

7 Source: Bureau of Indian Standards, Water Quality Standard, Drinking Water – Specification, IS 10500:1991 and India Standard (2005), Drinking Water – Specification (Second revision of IS 10500), ICS No.13.060.20.

Fluoride levels in drinking water had been found to range from 0.2 to 8.32 mg/L in Bihar, 1.5–18 mg/L in Gujarat, and 0.5–7.2 mg/L in Andhra Pradesh against the permissible limit of 1.5 mg/L.[6] Drinking water with fluoride beyond permissible levels leads to fluorosis which irreversibly affects teeth and bones. WHO has identified 17 (of the 32) states in India as "endemic" areas for fluorosis, severely afflicting 6 million people and placing another 66 million "at risk."

Arsenic causes nervous disorders, reduces IQ level in children, and in extreme cases can also cause cancer. Chromium is a known carcinogen. The presence of nitrates in drinking water leads to what is commonly called "blue baby disease," which hits infants and can lead to respiratory and digestive system problems. Table 9.2 provides an overview of the health impacts of a few common chemical contaminants in water, when present in excess.

In addition, 17% of households still need to fetch drinking water from a source located more than 500 m in rural areas or 100 m in urban centers, the burden being mostly borne by women.

9.4 THE EVOLVING POLICY FRAMEWORK AND PUBLIC INITIATIVES

In India, provision of safe drinking water is widely considered to be the responsibility of the government. The Indian Constitution confers government ownership over water resources, specifies water to be a state subject and gives citizens the right to potable water. At present, the states generally plan, design and execute water supply schemes through Public Health Engineering Departments, Rural Development Departments, or State Water Boards.

The central government initially focused on identification of "problem villages" (through the National Health Program in 1954), obtained technical support from UNICEF (in drilling of wells since 1969), and created a national program to address these challenges (through the Accelerated Rural Water Supply Program and, later, a more comprehensive Minimum Needs Program).

The National Drinking Water Mission was set up in 1986 (renamed Rajiv Gandhi National Drinking Water Mission in 1991). The first National Water Policy (1987) stated that national, rather than state or regional, perspectives will govern water resources planning and development and

[6] Source: Fluorosis Research & Rural Development Foundation. http://www.fluorideandfluorosis.com/fluorosis/prevalence.html (accessed 15.09.13.).

drinking water will have first priority while planning multipurpose water supply schemes. The 73rd Constitutional Amendments devolved responsibility for 29 subjects, including water supply, to local bodies (*Panchayati Raj Institutions* or PRIs) in 1992.

Community-based management of rural water supply was introduced through the Sector Reforms Pilot Projects in 1999. A Department for Drinking Water Supply (DDWS) was created (now the Ministry of Drinking Water and Sanitation). Efforts were enhanced with the adoption of MDGs in 2000 and, between 1990 and 2010, an estimated 522 million people gained access to improved water sources—sources that are, by design or active intervention, protected against external contamination from various local activities, including sewage seepage (UNICEF; WHO).

The new National Rural Drinking Water Programme (NRDWP), initiated in April 2009, introduced three major changes in water-service delivery: (1) changing norms of coverage from habitation to household (not reflecting in government statistics yet), with major implications on equity and social inclusion; (2) focus on multiple water sources, including traditional sources (such as roof rainwater harvesting systems); and (3) decentralized implementation, moving the onus of water supply provision from government departments to local elected bodies (or Panchayati Raj Institutions)—such as the Gram Panchayats and their subcommittees, Village Water Supply Committees (VWSCs)—and rural communities, explicitly endorsing the spirit of the 73rd Constitutional Amendment. A national program (*Swajaldhara*) for community-based drinking water supply was initiated in 2002 (James, 2011).

Under the new demand-driven/decentralized approach, members of the community are being consulted and trained, and users agree upfront to pay a tariff that is set at a level sufficient to cover operation and maintenance costs. It also includes measures to promote sanitation and improve hygiene behavior. However, in a study of *Swajaldhara* implementation in 10 states (Misra, 2008a), it was found that only about 10% of rural water schemes in India used a demand-driven approach, even though when used such an approach resulted in lower capital and administrative costs and better service quality compared to the supply-driven approach. However, since users have to pay lower or no tariffs under the supply-driven approach, this discourages them from opting for a demand-driven approach, even if such an approach is more likely to result in long-term sustainability.

While the policy thrust is toward decentralized management of water supplies, the investment focus continues to be on piped drinking water schemes that are often operated by state agencies. External planning and management, supported by planned government budgets, have led to a widening gap between the local needs and the quality of available services. Despite the strong push from central government, the states decide whether, and how, to give the responsibility to the PRI in rural areas or municipalities in urban areas, called Urban Local Bodies (ULB).

The Government of India (GoI) has spent INR 1105 billion (US$18 billion) up to 2008 on water projects (Khurana and Sen, 2008). The GoI's strategic plans for 2011–22 aim to ensure that 55% of rural households are provided with piped water supply by 2017 and it targets to increase the access to piped water supply to 90% by 2022 (GoI, 2011). Since 2009, NRDWP has allocated INR 764 billion (US$ 12.24 billion), with an average of INR 191 billion (US$ 3.05 billion) each year, although actual release, and utilization, has been lower.

9.5 CHALLENGE OF SUSTAINING SERVICES AND ENSURING QUALITY

The proportion of rural households served through piped systems has increased from 24.3% in 2001 to 30.8% in 2011, with 13 states out of a total of 35 states and union territories below the national average (Fig. 9.2). Moreover, only 17.9% of households with access to piped water in rural areas had access to treated water (Fig. 9.3).

Moreover, while about 32% of the economically privileged groups have piped connections on their premises, only about 1% of the underserved population has this facility.[8] The Census data reveal that while access to taps increased from 36.7% in 2001 to 43.5% in 2011, over a quarter did not incorporate any form of water treatment.

9.5.1 Is Universal Piped Water Supply a Pipe-Dream?

Only about 31% of households in rural India have access to tap water. The situation in central and eastern states of Maharashtra, Bihar, Jharkhand, Chhattisgarh, and Odisha is far more drastic, with none of the states having more than 10% of rural households with access to tap water (Fig. 9.4).

While many pilot WASH projects have been completed, solving the challenges to sustainability has been elusive—the World Health Organization

[8] Analyzed by UNICEF in 2010 based on data from National Family Welfare and Health Surveys in 1993, 1999, and 2006.

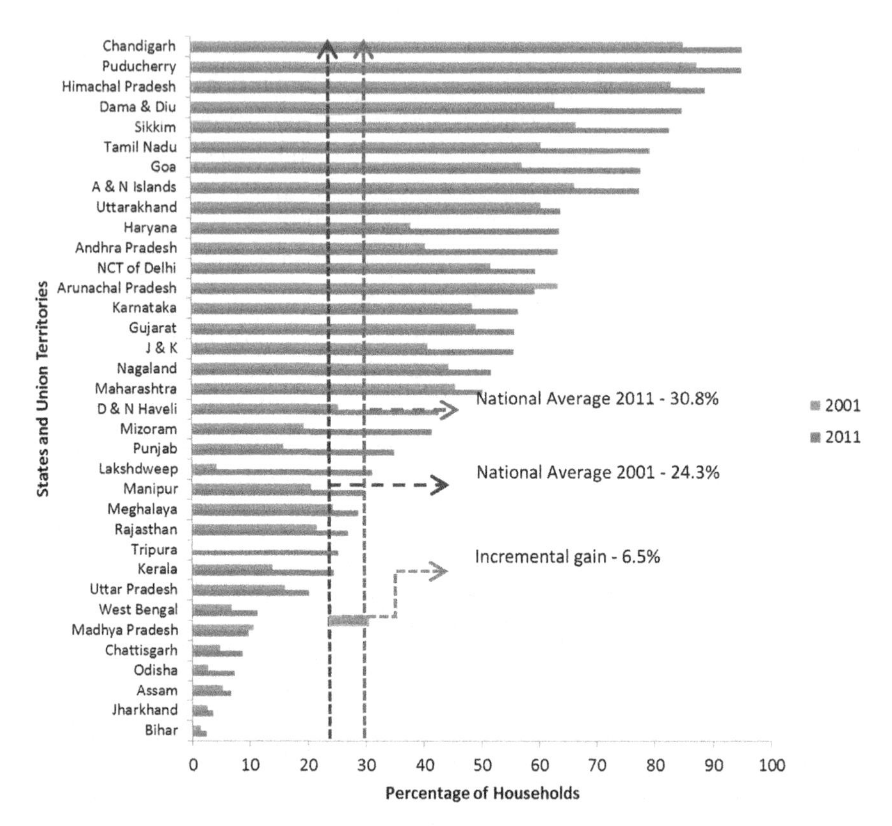

Figure 9.2 Percentage of rural households with access to piped connection by state in 2001 and 2011. *Source: Census, 2011.*

reports that in most developing countries 30–60% of rural water systems are inoperative at any given time (Brikké and Bredero, 2003).

Slippage is one of the main bottlenecks of achieving full coverage of water services in India. While data reported in 2013 show that India achieved almost 73% coverage in terms of improved water sources (piped and unpiped),[9] service delivery is not sustained. Thirty percent of habitations[10] were reported to have "slipped back"[11] as of April 2010, with dysfunctional and/or derelict infrastructure.

[9] Source: http://indiawater.gov.in/imisreports/Reports/Profile/rpt_StateProfile.aspx?Rep=0.

[10] "Habitations" are subvillage-level hamlets, often organized along caste lines, and hence tracked at that level for critical government functions and activities—including clean water availability, rural development, and education. There are approximately 1.6 million habitations in India. In some states, most villages have a single habitation; in others (notably Kerala and Tripura) there is a high ratio of habitations to villages.

[11] It is defined as the percentage of fully covered habitations that slip back to partial or nil coverage.

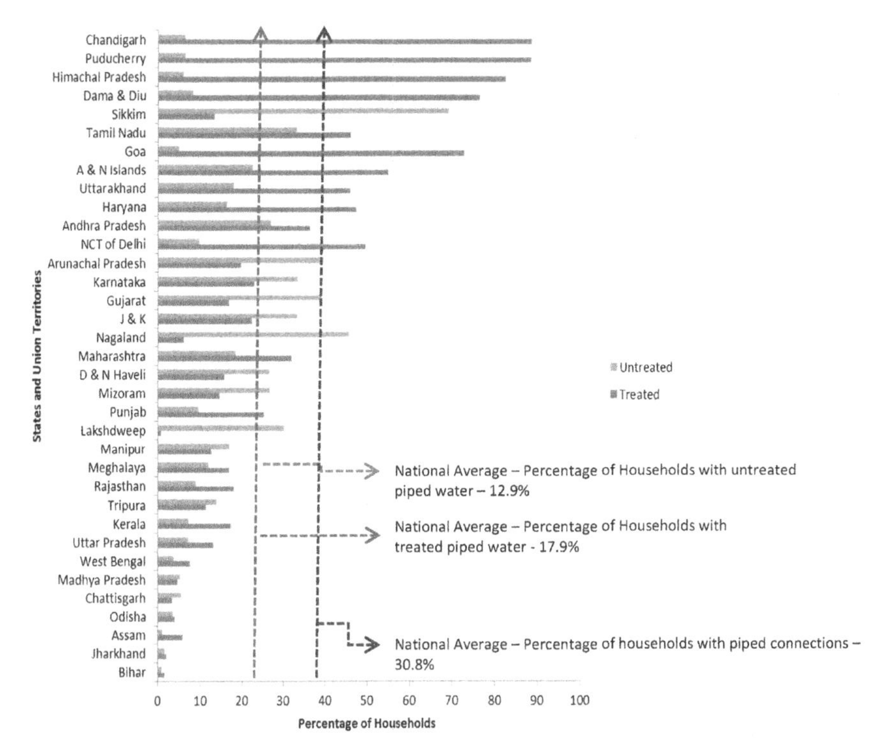

Figure 9.3 Percentage of rural households that received treated and untreated sources of piped water by states in 2011. *Source: Census of India, 2011.*

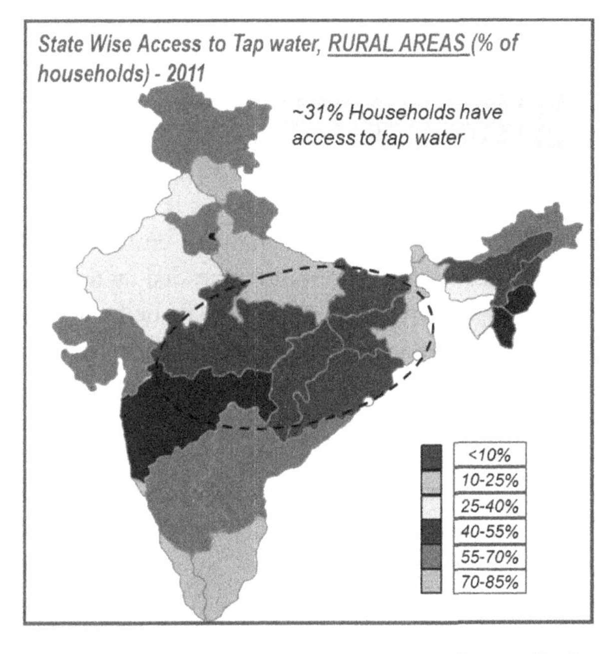

Figure 9.4 State-wise access to tap water. *Source: Census of India, 2011.*

The main reasons identified for slippage are drying -up of sources, sources becoming contaminated, lowering of yield of the source or seasonal shortage of water, lower per capita availability due to increase in size of population, poor operation and maintenance (O&M), and life of source outlived (Reddy et al., 2010).

Fourteen Indian States record slippage at rates higher than the national average (Reddy et al., 2010). The "Strategic Plan 2011–22" compiled by the Department of Drinking Water and Sanitation found that in 2010 out of 1.66 million habitations in India, 30% (0.49 million) had slipped back to partial coverage and a further 9% (0.14 million) saw their water quality affected negatively (GoI, 2011).

More than infrastructure, the real challenge lies in providing, and sustaining, good-quality services. Whether in small towns or megacities, or in single- or multivillage schemes, water is seldom supplied through a distribution network for more than a few hours a day, regardless of the quantity available. Most urban operations and all rural schemes survive on large operating subsidies and capital grants provided by the states. User charges are very low, but the cost of the alternatives on which users must rely far exceeds the full cost of providing a good-quality service. And while the poor may be the intended beneficiaries of the low user charges, they suffer most from the poor quality of service.

9.5.2 Tariff, Cost Recovery, and Willingness to Pay

The regulations under the Framework for Implementation 2009–12 by the Ministry of Drinking Water and Sanitation state that many States allow NGOs, private foundations and the private sector to provide drinking water at affordable prices. However, the regulations do not define the term "affordable price." The pricing of water in rural set ups is left to be decided by the local government (GoI, 2009). The Framework emphasizes that the state governments should constitute a committee headed by the Chief Secretary to look into setting up independent regulatory agencies, which could oversee pricing and terms of engagement between the bulk water utility and local governments. However, in the actual scenario, no proper committees are formalized, which leads to an ownership issue and absence of a proper mechanism to set the price for water.

Major financial institutions like the World Bank promote water pricing as a means for public water utilities to manage the allocation of existing water supplies more effectively (Dinar and Subramanian, 1997). There is limited published information on the actual price charged for the supply of drinking water in rural areas in India. A study by the World Bank in rural north Kerala showed that water supply tariffs in 1991 were INR 1 (US$0.016) per 1000L and were raised in 1993 to INR 1.50 (US$0.24) and further to INR 1.70 (US$0.27) in 1994. However, more recent data are not available.

The World Bank conducted a study in 2008 to understand WTP for improved water supply by rural household in India (Misra, 2008b). Rural households using private connections were found to be willing to pay about INR 60 (US$1.4) per month for improved water supplies and contribute an average of Rs 500–850 (US$11–19) toward the capital costs. Households using stand posts (one or more taps supplying water to many users) of piped water supply schemes were in general willing to pay about INR 20 (US$0.5) per month, with Rs 400–700 (US$9–16) for capital costs. The households using handpump schemes wanted better maintenance of the existing public handpumps and were willing to contribute about INR 6 (US$0.1) per month. WTP varied across the states with Maharashtra and Punjab showing the highest WTP, and Uttar Pradesh and Uttarakhand the lowest.

A microlevel research conducted in Akulam village, Kerala (Littlefair, 1998), revealed that households have different willingness to pay, depending on the nature of agency supplying water, ie, public water utilities or non-public water utilities. The study reported that the failures of the public water supply systems had reduced the faith of the villagers in the ability of the public water utility to provide a continuous and reliable supply of water. As a result, the price of water and the option of a more modern source of water were no longer household priorities. Households were willing to pay more money, irrespective of the income levels, to the non-public water utilities, indicating an elasticity of demand for a continuous supply of water. A middle-income-status villager was willing to pay a high proportion (75%) of his annual salary to dig a tube well, but would only pay 10% for a piped supply. A villager with a greater income status paid 35% of his household annual income to provide a private well but was unwilling to pay anything toward the public water utility supply system.

These studies determine WTP for improved water supplies, ie, continuous and reliable water provided to each household. No studies are available in rural India on household's WTP for safe drinking water. A review of experimental evidence, including price randomization techniques (that reveal valuation through real purchase decisions), contingent valuation techniques (that reveal preferences in a hypothetical market), and discrete choice analysis clearly revealed a willingness to pay that is unable to cover the full cost of technology and delivery of safe drinking water (Null et al., 2012).

9.6 THE POSSIBLE COMMERCIAL SOLUTION

In urban India, coverage levels of water supplied by public utilities are officially estimated at 90%, the gap spawning commercial ventures. Tanker services in urban areas, particularly in water-scarce regions, and investments in shallow-bore handpumps in many rural households, particularly in the Ganga basin, are examples.

There are challenges of poor quality of access, low reliability of supply, poor water quality, high distribution losses, and low cost recovery for which public–private partnerships (PPPs) are being introduced in many places. A study by WSP (2011) concludes that not only have the number of formal contracts with private sector participants increased since 1990, but variety and complexity are being introduced progressively in such contracts.

Several business models and projects have been initiated in India, driven primarily by three important policies, namely, National Water Policy 2002,[12] the Jawaharlal Nehru National Urban Renewal Mission (JNNURM), and the National Water Mission. It is estimated (Ernst & Young, 2011) that the Indian water sector has an investment potential of around USD 130 billion by 2030. Business opportunities revolve around the four key themes, namely, water demand management, water supply management, water infrastructure upgrading, and management of water utilities.

Inefficient utilities and rapid urbanization have forced individual households to address water challenges themselves. In urban areas, thanks to higher awareness and income levels, many households are increasingly opting for bottled water and/or household water treatment devices, and this surge in demand is being met by a growing market. In terms of revenue, India is the fifth-largest consumer of household water treatment (HWT) products in the world, after the United States, Japan, South Korea, and China (PATH, 2008). In India, annual unit sales of household water treatment products grew threefold between 1995 and 2005, to almost 3 million units (Baytel Associates, 2007). The per capita bottled water consumption in India is less than 5 L a year as compared to the global average of 24 L; however, consumption has risen rapidly in recent times. Today expenditures on bottled water in India total Rs 800 crore (about US $ 16 million) annually and this figure is growing at 25% per year, higher than any other country in

[12] The National Water Policy 2012 is now available.

the world (Vousvouras and Heierli, 2010). The city of Chennai alone is reported to have over 600 brands of packaged drinking water.

In rural and other underserved parts of India, the issue of safe drinking water is more complex because of (1) low density of scattered population, many living on less than a dollar a day; (2) overwhelming dependence on increasingly contaminated groundwater; and (3) extremely poor awareness and education on waterborne diseases and their debilitating impacts. Individual households address poor water quality from public services predominantly by boiling, and some form of straining and filtering (Rosa and Clasen, 2009).

Several new models for providing safe water are emerging in low-resource settings (PATH, 2008): (1) microfinance (MFI)-based financing; (2) a cooperative or self-help group (SHG) model for distributing health-related products; and (3) commercial community water systems, selling treated water to communities at a price that is far less than bottled water but sufficient to sustain the system.

9.6.1 Commercial Models in Community Water Systems

The demand for water and wastewater treatment in India is expected to grow at around 18–20% per annum, driven primarily by government contracts in cities.[13] In addition to municipal bodies, power, textiles, petrochemical, refineries, steel, pharma, fertilizers, and chemicals are the major users, and are driven by the increasing need to conserve and improve water use efficiency. The market is expected to reach around INR 22,000 crore (USD 4230 million) by 2018 (Infraline Energy, 2013).

With indigenous manufacturing of components and spares for various treatment technologies increasing over time, a large number of small and medium companies have emerged across the country that provide specific solutions. A few of these have focused on water treatment and supplies for rural drinking water needs. However, water treatment is regarded as a need for specific rural water supply schemes in India, and private companies are engaged by the government rather selectively.

9.6.1.1 Public–Private Partnerships (PPPs)

Many state governments have engaged local contractors for water treatment components of their schemes, while some have engaged only local

[13] Among the 250-odd companies operating in this segment, the following are the key—Thermax, Ion Exchange, VA Tech Wabag, Doshion, Driplex, Hindustan Dorr Oliver, Degremont, Praj, Aquatech, Paramount Ltd, Aquatech, Siemens, GE, UEM, Fontus, Veolia water, Bestech, Aqua Filsep, Permionics, Manas Water Tech, and Fontus Water.

mechanics, often on a part-time basis. The process was streamlined when public–private partnerships (PPPs) for the social sector, including water supply and sanitation, was adopted through the Eleventh Five Year Plan (2007–12). PPP projects are generally implemented through a competitive bidding process where the selected private operator develops and maintains the water treatment facility at a selected location. Usually these take the form of construction and management contracts of the following kinds:

1. Build–Operate–Transfer (BOT): The public sector entity—generally an agency of the state government—funds the construction, along with the operations and maintenance (O&M) for a specified period, usually 3–7 years. The private sector entity builds infrastructure matching tendered specifications and operates the systems for a specified period, beyond which it is generally handed over to the local Gram Panchayat. Water tariffs are set at a low level (generally INR 2–3 for 20 L). Rite Water, a private company operating in parts of central India, typically employs this model; and
2. Build–Own–Operate–Transfer (BOOT): Similar to the BOT model, but usually longer term (5–15 years), during which the assets are owned by the private entities, enabling depreciation and encouraging private investments. The government usually provides viability-gap funding if the revenue is less than the O&M costs and the private entity shares some of the revenue with the government if it is higher than the O&M costs and an established profit margin. The tariffs are generally government-prescribed, and remain lower than the O&M costs. The assets and operations are transferred to the Gram Panchayat after the contractual period. Water Health International (WHI) (a pioneering rural water treatment company) and its initial partner, Naandi Foundation, have typically operated through this model. Both entities have also been able to successfully raise external finances from financial institutions and donor agencies.

9.6.1.2 Case Study: Punjab

Punjab is one of the states suffering from serious chemical contamination of groundwater. Excessive pumping, use of fertilizers, pesticides, and insecticides has significantly impacted the quality of groundwater in the state. For instance, the amount of nitrate in the groundwater rose from 0.5 mg/L in 1972 to 5 mg/L in 2012, ie, 10 times, in 4 decades. In addition, out of 7000 schemes implemented by the state's water supply department, water in 1166 schemes is contaminated with heavy metals such as arsenic, mercury, and uranium.

One of the early solutions to this issue emerged in the form of community-level RO plants which spread fast throughout the state. Currently, there are an estimated 1800 RO plants installed in Punjab by the state government. The typical model of establishing these plants is a PPP, wherein government pays ~Rs 10–15 lakhs for installation of the plant and shelter to a private company. The company also operates the plant for typically 7 years, during which it is allowed to charge 10 paise per liter (~USD 0.04 per 20 L) from the users to recover their operational cost.

9.6.2 Community-Managed Systems

These are typically non-government ventures in which the local community is the majority stakeholder and plays an active role in financing, installing, and operating the system. Capital funding is generally provided by the NGO, with the community contributing some proportion. The technology partner provides assistance in system setup and training the community to perform O&M, and recurring expenses are met from nominal charges to consumers. There are innumerable NGOs operating on this model, including many supported by corporate funds for social responsibility. Bala Vikasa, Water for People, and Safe Water Network build the infrastructure through a small community/Gram Panchayat contribution and a relatively large grant, usually from charities and government agencies. What sets them apart, however, is their focus on cost recoveries and community contribution, albeit to varying extents.

9.6.2.1 Case Study: Participation

Participation holds the key to success in this model and different approaches are adopted to achieve the same. Water for People strives for "Everyone Forever," and hence engages through the Gram Panchayat which is responsible for collecting a uniform tariff from all and delivering a fixed quantity of treated water to all households. While this approach has met with grand success in West Bengal with larger and empowered Gram Panchayats, it is yet to succeed in neighboring Bihar, with smaller and weak Gram Panchayats. Bala Vikasa strives to serve the majority of a community, determined through a photograph of the Gram Sabha meeting, by promising grants of up to 80% of the installation costs of a water treatment facility. Safe Water Network seeks to improve the number of households purchasing treated drinking water through extensive promotional campaigns.

9.6.3 Private Enterprises

These are businesses in which a private company or an entrepreneur fully funds and owns a venture, is fully responsible for providing the services associated with it and charges a price that consumers are willing to pay. All activities, such as funding, installation, operation, and maintenance of the system, are carried out by a private company or an entrepreneur.

Private enterprises can simplify operations through clear ownership and control over activities, providing efficient services that can be sustained through revenues generated. Since full cost recovery leads to unaffordable prices, informal entities have emerged which often provide water of suspicious quality. Informal entities also tend to be opportunistic, exploiting the distress at times of shortages and prioritizing other preferences of consumers such as chilled water available at the doorstep.

Formal entities are not only challenged by cost recovery but many constantly struggle to access capital for infrastructure. Most have a stated objective that is clearly beyond a profit motive and may be referred to as social entrepreneurs. Prominent examples include WHI, Naandi Community Water Services, Healthpoint Services, Sarvajal, Waterlife, and Spring Health. WHI typically funds infrastructure employing a debt:equity model (60:40) with equity share from private investors and the debt component from IFMR, ICICI, and IFC. Similarly, Healthpoint Services provides all the funding for the infrastructure or sources it through government or private institutions. Ownership and operations are typically transferred to the communities after 10–15 years.

Sarvajal operates a franchise model where it owns the infrastructure and leases it to a local entrepreneur (the franchisee) for operations. Sarvajal invests in equipment, testing, education, and maintenance services, while the local entrepreneur typically pays INR 50,000 (US$800) as initial charges. The revenues are shared 60:40 between the entrepreneur and *Sarvajal* (Macomber and Srivastava, 2011).

Waterlife and Spring Health employ the entrepreneur model where a local entrepreneur shortlisted by community elders is selected to operate as a local partner. In the case of Spring Health, this entrepreneur, usually a local retailer, pays a one-time joining fee of INR 5000 (US$80), which is one-fifth of the cost of the chlorination equipment. Waterlife employs more costly RO infrastructure and hence relies more on the distribution network of its entrepreneurs.

9.6.3.1 *Case Study: The Market at* Jattari

Jattari is a small town and a Nagar Panchayat in Aligarh district of Uttar Pradesh (28°1′16″N; 77°39′31″E), with a population of about 20,000. While its economic base is agriculture, a range of small-scale commercial establishments seek to provide urban goods and services, taking advantage of the National Highway 22A, to a rapidly urbanizing community. *Jattari* is about 20 km from the Yamuna river but has brackish groundwater (salinity between 900 and 1300 ppm), which is the regular source of drinking water in the region.

In December 2008, Bharat Pratap Singh invested over Rs 9 lakh to set up a four-stage (sand, carbon, micron, and membrane) treatment plant of 1000 L/h capacity to supply "Mountain Valley" brand of water. In November 2009, *Chandrapal Sharma* (the son of a local teacher) teamed up with *Sushil* (whose family had established itself in the business of sweets and condiments), to invest Rs 6–7 lakhs to set up another plant of identical capacity to supply the "Ganga Neer" brand of water. The plants were assembled in the nearby towns of Faridabad and Ballavgarh, respectively, and the dealers were committed to after-sales services.

The production facilities of *Ganga Neer* were crammed into a small, dark room with a fenced courtyard in front, on a small alley within the interior of the town. *Sushil* and *Chandrapal* lived on the other side of the alley. They operate the plant, oversee the distribution and, with the help of a laborer, are able to cater to a peak demand of about 300 consumers. *Bharat Pratap*, on the other hand, decided to lease his plant after a year to *Praveen Kumar*, who paid Rs 12,000 per month as rent. *Praveen* works hard, operates the plant, distributes the water, and maintains daily records, successfully catering to a peak summer demand of about 110 consumers.

The demand for chilled water, particularly at the commercial establishments in this small town, emerged with the worsening power situation within the next couple of years. The *Ganga Neer* unit purchased chilling units with a total capacity of 1000 L per hour. Mountain Valley unit followed suit. In fact, it also offered free containers (known as *Mayur* jugs, after the popular brand), and then *Ganga Neer* followed suit. These containers were refilled on demand, through 20-L bubble-top cans ferried rapidly on cycle rickshaw and motorized rickshaw, making the business labor-intensive. Hiring and retaining laborers became the most pressing problem for the entrepreneur, and cheap motorized vehicles were introduced.

Jattari receives about 6–10 h of power supply, mostly between late night and early morning. The household refrigerators can preserve food overnight but are unable to keep water cooled during the waking hours. While this creates demand for chilled water, the chilling units themselves require about 5 h of uninterrupted power supply. Both the ventures now have to invest in diesel-powered electricity generators which consume an average of 2 L of diesel per hour.

The demand for treated water is inelastic to price in rural areas. Conspicuous value addition, like doorstep delivery or chilling, does help in raising the price, but also shrinks consumption. Besides, the market itself shifts from producing treated drinking water to an aspirational beverage. It also tends to restrict and alter the space for social entrepreneurs that are focused only on providing treated drinking water to rural communities. Facilitative rules, massive awareness campaigns, and supportive monitoring infrastructure can correct such market distortions.

9.7 SUSTAINABILITY OF TREATED DRINKING WATER SYSTEMS

9.7.1 Governance and Decentralised Management

Sustainability of drinking water sources and systems are major issues in the rural water supply (RWS) sector (GoI, 2009). Allocation of dependable and reliable water supplies for drinking purposes in every habitation appears to be the first step. However, this requires knowledge and control over available water resources, something that is not only missing but is becoming complex due to emerging multiplicity of demands and increasing uncertainties induced by climate variability. One successful example pertains to a unique community-managed system, implemented in partnership with the state government, the Water and Sanitation Management Organization (WASMO) in Gujarat. WASMO is active in every district of Gujarat and provides water supply services in about 15,000 out of a total of 18,000 villages in the state. The water supply services are governed by a tripartite agreement signed between WASMO, the local *Gram Panchayat* and the *Pani Samiti* (Lockwood and Smits, 2011).

However, a water supply service is sustainable when the source is dependable; the system functions properly; and is able to deliver the appropriate level of benefits, ie, quality, quantity, convenience, and reliability to consumers throughout its design life (Sara and Katz, 2005). At the same time the costs for operation, maintenance, administration, and equipment replacement are covered at the local level with limited but feasible external support (Brikke, 2002). Sustainability of small, distributed water supply services depends on suitable ownership, robust governance, selection of the correct technology, and financial viability.

Ownership of the water system is a key factor. Poor ownership of water supply systems and sources by rural communities and poor O&M have been shown to be the two key factors that have contributed to rapid deterioration of the water supply facilities resulting in non-availability of the designed services (GoI, 2009). Community ownership has been shown to ensure higher sustainability of services as against ownership by the government, NGOs, or social enterprises (Fielmua, 2011). In India, NRDWP promotes a decentralized approach through PRI and community involvement and the central government increasingly distributes responsibilities for the water supply system to the local level in order to ensure drinking water security to the rural community.

In order to ensure ownership, the community needs to understand the benefits of treated water. Community awareness, together with campaigns to promote the tangible health benefits of treated water for drinking water supply will contribute to increasing the demand. A study conducted by the Planning Commission in five states (Karnataka, Himachal Pradesh, Rajasthan, Assam, and West Bengal) indicates that 71% of the Gram Panchayats were part of various water awareness campaigns (Planning Commission, 2010). However, it is important that the benefits of these awareness programs percolate down to the rural households to ensure full involvement of the local communities.

Sustainable governance entails a tailored structure to manage, operate, and maintain the water systems supplying treated water. A governance structure which involves all the stakeholders will demonstrate a business ability to consider and minimize potential shortcomings, enhance responsibilities, and ensure its ongoing operation.[14] In the case of public–private partnerships, selected members of all financing bodies, both public and private, will constitute the governance structure. All functions, including water sourcing, operations, maintenance, and distribution, will be managed by committees designated for the purpose. Separate committees for finance, public relations, and technology support and training might also be constituted. The governance structure will be headed by a chairman and all entities will report to this person (PWC, 2005). However, it remains bound by government policy in pricing, ownership, geography, etc., which may inhibit the flexibility of the plant operator to enhance the program's viability and effectiveness.

A better alternative is to raise the governance a step further, as in the case of WASMO in Gujarat, and introduce flexibility to experiment with different models of integrating water treatment facilities into rural drinking water supply schemes. The treatment facilities could then be either managed as an integral part of the water supply system through the Gram Panchayat and the *Pani*

[14] Source: http://www.goodbusinessregister.com.au/fact_sheets/GBR_Tips_for_SMEs_sheet.pdf.

Samiti, outsourced to a local entrepreneur for its sustainable management or operated through a contractual partnership with a social enterprise.

9.7.2 Key Issues in Decentralized Management

The technology employed for water treatment needs to be appropriate for the water quality problems facing a habitation. In addition, the diffusion of any technology is also dependent on its ease of maintenance and acceptance of consumers. Reverse osmosis, ultraviolet treatment, and ozonization are the most effective for a wide range of water quality challenges, but their feasibility in a rural community setting needs to be evaluated in terms of capital expenditure, impact on the local environment, societal acceptance, and skilled human resources for operating and maintaining the systems. Simple disinfection (chemical oxidation) and basic filtration, where effective, can be cheap and efficient solutions.

In rural areas of India, the technical expertise and management skills that are required to manage the technically complex rural water supply systems are seldom present. Appropriate training at each level across the value chain of the water supply sector is needed to ensure an efficient operation and management of the water treatment units. Technical training is usually provided through training kits and practical demonstrations by skilled technicians brought from outside the community. While basic maintenance and simple repair become a local responsibility, a network of backup support needs to be created for spares and complex repairs. Managerial training, critical to sustainable operations, is often missing or diluted. This would typically include consumer relations and education, accounting procedures, and environment management. Operating manuals, exposure visits to best practices, and formal training sessions are some of the tools for effective management training.

Financial viability is a key to sustainability, achieved only when all the capital and operating costs are recovered from consistent revenues. The key challenges include wide disparities in affordability, uneven levels of consumer awareness, and additional costs in transportation and delivery. A viable financial proposition must be developed that reduces costs for poor consumers while ensuring revenues that are adequate for sustaining operations and scaling up of the system. A partnership model shares risks and responsibilities while a proprietary model ensures accountability and efficiency.

Environmental sustainability depends on decisions made on the basis of up-to-date knowledge and information, institutional mechanisms to debate and discuss options for optimal use and management of waste, and innovative ideas that integrate and explore multiple use, promote options for appropriate reuse and recycling and introduce technologies.

9.7.3 Emerging Market and Opportunities

Private enterprises generate local resources, respond to consumer needs through timely investments, and build a value-chain that enables sustainability of services. At the moment, such enterprises serve at least 7000 rural and peri-urban communities across India, and are set to grow rapidly in the near future (Safe Water Network, 2014). Application of RO technology in small-scale, community-level systems is widely considered to be the best solution for treatment of raw water for drinking purposes (Macomber and Srivastava, 2011). Expensive technologies like RO, ultrafiltration, and ion-exchange enhance consumer confidence, particularly in a country affected by a broad range of contaminants in groundwater-based drinking water sources. However, adoption of private enterprise models reveals an interesting pattern that correlates with household access to tap water. Rural areas with more than 25% of households having access to tap water, roughly coinciding with the region outside the ellipse in Figure 9.4, appear to have a higher willingness to pay for treated drinking water. Major enterprises operating in this region include pioneers such as WaterHealth International (WHI), Naandi Foundation (now Naandi Community Water Services), and Sarvajal. Growth of these enterprises has catalyzed the development of a supply chain for technology components and human resources, factors that increase reliability, help drive down costs and achieve scale.

Private enterprises operating in areas with less than 10% of households having access to tap water, roughly the region within the ellipse in Figure 9.4, employ a range of models to address lower willingness of consumers to pay for water. While Rite Water chiefly depends on government funding through contracts, Water for People partners Gram Panchayats to ensure payments. Waterlife seeks to build livelihood dependencies through a network of water vendors, while Spring Health adopts low-cost disinfection technology to maintain profitability in its distribution chain.

In addition to financial viability of private enterprises, there are concerns of social equity and environment impact. Government funding and grant-based non-profit entities ensure a low market price for treated drinking water. However, it is unclear whether such implicit (and explicit) subsidies actually lead to increased adoption of treated drinking water.

Typically, environmental concerns are also ignored, like in many other sunrise sectors. However, though not yet evident, social and environmental concerns are likely to be addressed progressively through an emerging trend of bundled services. Healthpoint Services India combines primary healthcare, basic diagnostics, and telemedicine along with safe drinking water, while Sulabh International has initiated vending of safe drinking water alongside its robust network of public toilets in India. Such bundling of WASH services provides opportunities for differential pricing that align with the use of waters of different quality.

REFERENCES

Ali, M., Lopez, A.L., You, Y.A., Kim, Y.E., Sah, B., Maskery, B., Clemens, J., 2012. The global burden of cholera. Bull World Health Organization 90 (3), 209–218A.

Baytel Associates, 2007. The Indian Market for Home Water Treatment Products. Beytel Associates, San Francisco, USA.

Brikke, F., 2002. Operation and Maintenance of Rural Water Supply and Sanitation Systems: A Training Package for Managers and Planners. World Health Organization, Geneva.

Brikké, F., Bredero, M., 2003. Linking Technology Choice With Operation and Maintenance in the Context of Community Water Supply and Sanitation. World Health Organization and IRC Water and Sanitation Centre, Geneva.

CGWB, 2010. Groundwater quality in shallow aquifers of India. Central Groundwater Board. Ministry of Water Resources. Government of India 117.

Dinar, A., Subramanian, A., 1997. Water Pricing Experiences: An International Perspective. The World Bank, Washington, DC.

Ernst & Young, 2011. Water Sector in India – Emerging Investment Opportunities. Ernst & Young, New Delhi.

Fielmua, N., 2011. The role of community ownership and management strategy towards sustainable access to water in Ghana (a case of Nadowli district). Journal of Sustainable Development 4 (3), 174–184.

Government of India (GoI), 2009. Movement Towards Ensuring People's Drinking Water Security in Rural India – Framework for Implementation 2009–2012. Rajiv Gandhi National Drinking Water Mission, Department of Drinking Water Supply, Ministry of Rural Development, Government of India, New Delhi.

Government of India (GoI), 2011. Strategic Plan – 2011–2022: Ensuring Drinking Water Security in Rural India. Department of Drinking Water and Sanitation, Ministry of Rural Development, Government of India, New Delhi.

Health Status Indicators, 2010. The Status of Health and Education in India: Critical Questions in the Nation's Development.

Infraline Energy, 2013. Water and wastewater treatment market in India. Infraline Energy. Market research report series. New Delhi.

James, A.J., 2011. India: Lessons for Rural Water Supply; Assessing Progress Towards Sustainable Service Delivery. IRC International Water and Sanitation Centre and iMaCS, The Hague, The Netherlands; New Delhi.

Kumar, M.D., Shah, T., 2004. Groundwater contamination in India: The emerging challenge. Hindu Survey of the Environment 2004, 7–12.

Khurana, I., Sen, R., 2008. Drinking Water Quality in Rural India: Issues and Approaches. WaterAid India, New Delhi.

Littlefair, K., 1998. Willingness to Pay for Water at the Household Level: Individual Financial Responsibility for Water Consumption. MEWEREW Occasional Paper No. 26. School of Oriental and African Studies, University of London, London.
Lockwood, H., Smits, S., 2011. Supporting Rural Water Supply: Moving Towards a Service Delivery Approach. Practical Action Publishing, UK.
Macomber, J., Srivastava, M., 2011. Sarvajal: Water for All. Harvard Business School, USA, Case Study No. N9-211-028, dated Feb 9, 2011; 29 pp.
Misra, S., 2008a. Inefficiency of Rural Water Supply Schemes in India. The World Bank, Washington, DC.
Misra, S., 2008b. Review of Effectiveness of Rural Water Supply Schemes in India. The World Bank, Washington, DC.
Null, C., Michael, K., Edward, M., Hombrados, G., Meeks, J., Zwane, R., Peterson, A., 2012. Willingness to pay for cleaner water in less developed countries: systematic review of experimental evidence. International Initiative for Impact Evaluation. Systematic Review 006, 46.
Planning Commission, 2010. Evaluation Study on Rajiv Gandhi Drinking Water Mission. Programme Evaluation Organization, Planning Commission, New Delhi.
Program for Appropriate Technology in Health (PATH), 2008. Activities and Stakeholders in the Global Water Sector: A Preliminary Analysis. PATH, Seattle, USA.
Prüss-Üstün, A., Bos, R., Gore, F., Bartram, J., 2008. Safer Water, Better Health: Costs, Benefits and Sustainability of Interventions to Protect and Promote Health. World Health Organization, Geneva.
PWC, 2005. Corporate Governance Toolkit for Small and Medium Enterprises, second ed. PWC.
Reddy, V.R., Rao, M.R., Venkataswamy, M., 2010. Slippage': The Bane of Rural Drinking Water Sector (A Study of Extent and Causes in Andhra Pradesh). Microeconomics Working Papers No. 22734. East Asian Bureau of Economic Research.
Rosa, G., Clasen, T., 2009. The global prevalence of boiling as a means of treating water in the home. In: Brown, J., Outlaw, T., Clasen, T., Wu, J., Sobsey, M. (Eds.), Safe Water for All: Harnessing the Private Sector to Reach the Underserved. International Finance Corporation/World Bank, Washington, DC.
Safe Water Network, 2014. Community Safe Water Solutions: India Sector Review. Safe Water Network, NY, p. 68.
Sara, J., Katz, T., 2005. Making Rural Water Supply Sustainable. Report on the Impact of Project Rules. UNDP-World Bank Water Sanitation Programme, Washington, DC.
Srikanth, R., 2009. Challenges of sustainable water quality management in rural India. Current Science 97 (3), 317–325.
UN India, 2015. India and the MDGs: towards a sustainable future for all. United Nations, p. 27.
United Nations, 2015. The Millennium Development Goals Report 2015. United Nations, New York, p. 72.
Vousvouras, C.A., Heierli, U., 2010. Safe Water at the Base of the Pyramid; How to Involve Private Initiatives in Safe Water Solutions. 300in6 Initiative.
Waddington, H., Snilstveit, B., White, H., Fewtrell, l, 2009. Water, sanitation and hygiene interventions to combat childhood diarrhoea in developing countries. International Initiative for Impact Evaluation. Synthetic Review 001, 119.
Water and Sanitation Program (WSP), 2011. Trends in Private Sector Participation in the Indian Water Sector: A Critical Review. The World Bank, Washington, DC.

CHAPTER 10

Positive Externalities of Surface Irrigation on Farm Wells and Drinking Water Supplies in Large Water Systems: The Case of Sardar Sarovar Project

S. Jagadeesan
Former Managing Director, Sardar Sarovar Narmada Nigam Ltd, Gandhinagar, Gujarat, India

M.D. Kumar
Institute for Resource Analysis and Policy, Hyderabad, India

M.V.K. Sivamohan
Institute for Resource Analysis and Policy, Hyderabad, India

10.1 INTRODUCTION

Large irrigation projects are targets of increased criticism worldwide for the negative social and environmental impacts they are likely to cause. As noted by Biswas and Tortajada (2001), Verghese (2001), and Shah and Kumar (2008), critics argue that the costs outweigh the intended benefits. However, some scholars argue that developing countries have no other choice than to build large storages (Biswas, 1994; Biswas and Tortajada, 2001; Kumar, 2009). The issue, according to them, is how best to improve their overall effectiveness for human welfare, poverty eradication, and preservation of the environment and how to make beneficiaries of those who pay the costs of the construction of such large structures. The Sardar Sarovar Project (SSP) constructed on the Narmada River, one of the most controversial water development projects in independent India (Verghese, 2001; Shah and Kumar, 2008), is no alien to these criticisms.

The SSP is expected to irrigate nearly 1.8 million hectares of agricultural land in Gujarat through a network of canals comprising the main canal, branch canals, distributaries, minors and water courses, with a total length of 75,000 km. The entire canal system is designed for gravity irrigation and

Rural Water Systems for Multiple Uses and Livelihood Security
ISBN 978-0-12-804132-1
http://dx.doi.org/10.1016/B978-0-12-804132-1.00010-X

is already irrigating nearly 0.62 m ha of land (Alagh, 2010). The state-wide water grid, with the Sardar Sarovar Canal–based Drinking Water Supply Project as its core, is expected to bring in developmental benefits to around 47 million people in the state once it is completed (Biswas-Tortajada, 2014).

Until recently, the available analyses of performance of SSP (Talati and Shah, 2004; Parasuraman et al., 2010; Shah et al., 2009) are based on narrow objectives and with ideological overtones, and do not consider the indirect benefits. For instance, gravity irrigation produces several positive externalities on local groundwater regimes (Chakravorty and Umetsu, 2003; Shah and Kumar, 2008; Watt, 2008), through changes in water availability in wells in the command area for irrigation and availability and quality of groundwater for drinking and domestic uses, as well as water table conditions. These changes are affected through groundwater recharge from irrigation return flows and canal seepage (Shah and Kumar, 2008; Watt, 2008).

However, the analysis of the socioeconomic impacts of SSP, carried out by the Institute for Resource Analysis and Policy (IRAP) in 2012, took into account these indirect benefits (IRAP, 2012). The study examined the externalities of canal irrigation in terms of the following: (1) saving in cost of well deepening in the command area; (2) saving in energy cost of groundwater irrigation; (3) incremental economic surplus of well irrigators from crop and dairy production; (4) increase in wage rate of agricultural laborers; (5) reduction in cost of domestic water supplies in the command area villages and towns; and, (6) increase in land prices. The findings helped draw the positive externalities of canal irrigation on well irrigation and drinking water supplies.

10.2 POSITIVE EXTERNALITIES OF GRAVITY IRRIGATION FROM LARGE WATER SYSTEMS

The way irrigation efficiencies are analyzed by irrigation managers has changed dramatically in the last few decades (Molle and Turral, 2004). Based on the notion that the water applied in the field in excess of the crop water requirement is waste, in the past irrigation efficiencies were assessed by considering the total amount of water consumed by the crop (ET) in relation to the total amount of water applied in the field (Seckler, 1996; Allen et al., 1998). Given the fact that a significant proportion of the water applied in the field can actually get captured by downstream users including the form of well water available from irrigation return flows (Allen et al., 1998), such notions no longer help in analyzing the efficiency with which water from irrigation systems is used (Seckler, 1996).

The recharge benefits of gravity irrigation, through canal seepage and return flows from canal-irrigated fields, is not only well articulated (Dhawan, 2000), but also well established (Watt, 2008). The most recent evidence is from erstwhile Andhra Pradesh, where the average annual groundwater balance in blocks receiving canal irrigation was found to be far better than that in blocks outside canal commands. The proportion of blocks falling in the "overexploited" category was much less in the command areas as compared to non-command areas. An estimate by IRAP (2015) showed that the average rate of recharge in command area was 0.275 m against 0.10 m in non-command areas. This sharp difference was because of additional recharge from canal irrigation and reduced groundwater pumping owing to the presence of surface water through canals.

The recharge can improve the yield of agro-wells in the command area and increase the area irrigated. At the same time, rising water tables can reduce energy consumption for pumping groundwater, and reduce the incidence of well failures (Ranade and Kumar, 2004; Shah and Kumar, 2008; IRAP, 2012). Reduction in energy consumption for groundwater pumping and incidence of well failures has positive implications for the economics of irrigated agriculture. As pointed out by Chakravorty and Umetsu (2003), return flows from canal-irrigated fields and canal seepage can sometimes generate greater economic value than it would otherwise generate if used directly for irrigation.

The reason is that farmers using the recycled water through lifting devices will be able to use it in a more controlled fashion than those who are using surface irrigation, resulting in better application of water, higher water-use efficiency in terms of biomass produced per unit of ET (Kumar and van Dam, 2013). Researchers in the past have highlighted some inherent advantages of well irrigation over surface irrigation. Daines and Pawar (1987) suggest that well irrigation in India is significantly more productive than surface water irrigation. Similar data are available for Spain (Hernandez-Mora et al.). These productivity differences could be due to several factors: better understanding of the resource availability, enabling the farmers to plan their crops; greater access to and control over water enjoyed by well-owning farmers resulting in timely irrigation; and greater control over water delivery from wells resulting in better input use efficiencies and on-farm water management. Timeliness of water delivery and excess water deliveries had a significant impact on crop yield, with the impact of the first being positive and that of the second being negative, as shown by a study on irrigated rice production in Sone irrigation command in Bihar (Meinzen-Dick, 1995).

Studies carried out in the Indus basin in Pakistan and in the Murray–Darling basin in Australia show that irrigation return flows and seepage from unlined canals are high enough to improve the groundwater regime by raising the water levels in semiarid and arid regions with deep water-table conditions, if water is introduced for a long time period. Such changes happen due to reduced pumping of groundwater with the introduction of canal water, gradual increase in moisture storage in the unsaturated zone, which increase the unsaturated hydraulic conductivity of the soil media, and an increase in moisture pressure (hydraulic) gradient of the soil (Watt, 2008). While the first factor reduces or keeps the vertical distance for movement of recharge water, the second and third factors increase the rate of vertical movement of soil water (source: based on Richards Equation and van Genuchten Equation as cited in Watt, 2008, pp. 101–102). Hence, the so-called wastage from canal irrigation produces positive externalities on groundwater irrigation in a region.

Recharge from canal seepage and irrigation return flows can also induce positive externalities on drinking water supplies, though these positive externalities can be very location-specific. It can often reduce the concentration of minerals like fluorides and salinity in native groundwater, which causes serious health hazards (Ranade and Kumar, 2004) as canal water is mostly free from minerals. Continuous recharge from canals and irrigated fields can ensure good yields of drinking water sources, whose sustainability is threatened by aquifer depletion due to overpumping. The groundwater in the confined aquifers of the alluvial areas of Gujarat contains minerals such as fluoride and other salts. These make groundwater unfit for drinking in many pockets. The examples are districts of north Gujarat which record high TDS levels in groundwater affecting millions of people living in rural and urban areas of these districts, using the groundwater for domestic purposes, including drinking and cooking; and Sidhpur and Patan areas which record high fluoride levels in groundwater, affecting the lives of millions of people with serious health impacts in the form of dental and skeletal fluorosis (IRMA and UNICEF, 2001), and negative impact on girls' education. In the rural areas affected by high TDS and fluoride, water had to be supplied from regional water supply schemes which import water from distant reservoirs being prohibitively costly to the state exchequer.[1]

The large cities in Gujarat, which have contaminated groundwater in the aquifers underlying them, are now dependent on water supplied from

[1] The regional water supply scheme based on Dharoi reservoir, which was primarily meant for irrigation, is just one of the many examples of this growing trend.

Sardar Sarovar reservoir using the Narmada-Mahi pipeline scheme. Dilution of groundwater, occurring as a result of the seepage from Narmada canal and percolation from river beds, would ease the pressure on the regional water supply schemes, and that replaced water could be used for productive purposes. Improvements in groundwater quality can have positive impacts on drinking water in areas where poor-quality groundwater is used to meet domestic water demand. Higher well yields can better the drinking water situation in areas that rely on hard rock aquifers, where wells dry up during summer.

India heavily depends on groundwater to meet drinking water needs: 50% of urban water supply and 80% of rural water supply requirements are met from this resource. This includes areas with groundwater quality problems (Kumar, 2007). Often public water utilities relax chemical quality standards prescribed by the World Health Organization and Bureau of Indian Standards for drinking water, to ensure drinking water security in difficult areas. Removal of minerals from groundwater to make it potable is emerging as a major challenge for public water utilities in both rural and urban areas. As a result, the indirect impacts of augmented groundwater recharge from canals can extend far beyond the agricultural sector.

While some of these externalities add to private benefits, some of them reduce public costs. For instance, reduction in well failures and improved well yields bring down the production costs incurred by farmers, thereby raising the net private gain from irrigated crop production. Even when reduced energy consumption for groundwater pumping decreases, to an extent, overall economic costs, it does not necessarily affect the private costs for the farmers as they do not have to incur the marginal cost of electricity for pumping groundwater under the subsidy regime. Similarly, in the case of drinking water, improved performance of water supply sources reduces the public cost for providing this service. Such costs are significant in hard rock areas, when hundreds of thousands of hand pumps, open wells, and bore wells go dry during summer, and the utilities have to supply water to the rural and urban areas through tankers.

10.3 STUDY LOCATION, METHODS, AND DATA

The study was carried out in six districts of Gujarat receiving surface irrigation from the Sardar Sarovar Project, either directly through gravity or through water lifting from the irrigation canal. Ahmedabad and Mehsana receive water from the main system and Bharuch, Narmada, Vadodara, and

Table 10.1 Sampling size

Sr. No.	Parameter analyzed	No. of observation wells January (Winter)		No. of observation wells May (Pre-monsoon)		No. of observation wells October (Post-monsoon)	
		1996–99	**2005–10**	**1996–98**	**2004–09**	**1996–98**	**2004–09**
1	SWL	222	222	222	222	222	222
2	TDS	222	222	222	222	222	222

Source: Sardar Sarovar Narmada Nigam Ltd, Gandhinagar.

Panchmahals receive water from gravity canals. The study was based on primary data from well irrigators in the command area of the gravity irrigation system and well irrigators in the area receiving water from the main system through lifting. Data on the drinking water situation were also collected from villagers in the command area; and secondary data on groundwater-level trends and groundwater quality (TDS) in observation wells in the designated command area of SSP (Table 10.1) for two time periods, one pre-Narmada and one post-Narmada, were collected from SSNNL.

The primary data include information of both pre- and post-water supply from Narmada regarding water levels, yields, and well-deepening incidence in the command area; and water levels, yields, and well-failure incidence in villages under canal commands. It also comprises input and output data for farming enterprises, both crops and livestock, namely area under different irrigated crops; livestock composition and size; quantum and cost of various crop and livestock inputs and outputs.

For analyzing the regional-level impact on groundwater, the water level trends in the region, during two time periods, ie, prior to introduction of Narmada waters and post-Narmada water introduction were analyzed. Furthermore, the analysis covered comparing water levels in the observation wells, from January to January (winter), from May to May (pre-monsoon), and from October to October (post-monsoon). For instance, the change in water levels from January 1996 to January 1999 (pre-Narmada) is compared against the change in water levels in the same observation wells from January 2005 to January 2010; from May 2004 to May 2009; and from October 2004 to October 2009. In order to analyze the changes in groundwater quality, the differences in TDS (ppm) of groundwater samples collected from observation wells, for the same time periods as mentioned above, are analyzed.

For analyzing the geohydrological impacts, the static water levels (SWLs) in both time periods were taken from 222 observation wells. Similarly, the TDS values of water from these 222 observation wells were taken for analyzing the geohydrochemical impacts. The analysis was carried out for seven districts, namely, (1) Ahmedabad; (2) Baroda; (3) Banaskantha; (4) Bharuch; (5) Kheda; (6) Mehsana; and (7) Surendranagar.

10.4 CHANGES IN GROUNDWATER AVAILABILITY AND QUALITY IN THE SSP COMMAND AREA

The analysis of regional data on the groundwater levels showed that there was a marked improvement in the groundwater levels in the command area, after the introduction of Narmada canal water (2004–09) as compared to the pre-Narmada period of 1996–98. By and large, static water levels in all the districts except for Kheda not only showed an upward trend but also higher than pre-Narmada water levels. Unlike what is generally observed in surface irrigation commands, extensive groundwater use in the SSP command area arises from the conjunctive use of surface and groundwater. The rise in water levels has helped the farmers in the command area reduce energy requirement per unit of groundwater pumped and the cost of deepening wells, thereby resulting in their increased irrigation potential (IRAP, 2012).

As regards groundwater quality, the overall water quality trend during the period of Narmada water introduction is positive, with the average TDS values declining consistently over a period of time in most districts or the rate of increase in TDS values declining. Comparing this trend with the pre-Narmada trend shows the following outcomes. In districts where the TDS had reduced prior to Narmada, the "lowering trend" continued post-Narmada also, but with higher or lower annual rate of drop. The first type of situation was encountered in Ahmedabad for January 2005–10 and May 2005–10 as compared to January 1996–99 and May 1996–98. The second type of situation was encountered in Bharuch district irrespective of the season considered for analysis. In Baroda district, the same situation was encountered during January 2005–10 and May 2004–09 as against January 1996–99 and May 1996–99, respectively (IRAP, 2012).

In two districts, namely, Kheda and Ahmedabad, there has been a trend reversal, with the rising trend being replaced by a lowering trend. Only in one district, ie, Surendranagar, was the trend reversal negative. However, what is most interesting is that the trend in most districts is not consistent across seasons like what was found in the case of water-level trends. The two districts where the "positive trend" of higher rate of reduction in TDS values or annual increase in TDS getting replaced by annual reduction in TDS, was

Table 10.2 Reduction in depth to water level in wells in different seasons after the introduction of Narmada water in different districts (ft)

Season	Ahmadabad	Bharuch	Mehsana	Narmada	Panchmahals	Vadodara
Monsoon	16.39	6.39	47.34	13.16	14.47	14.63
Winter	15.31	7.94	48.74	13.21	16.32	13.68
Summer	53.00	6.82	33.15	9.82	18.41	30.03

Source: Authors' own analysis based on data from field survey.

consistent across seasons, are Ahmedabad and Kheda. Bharuch is the only district where the negative trend of lower rate of reduction in TDS was found to be consistent irrespective of the season chosen for analysis (IRAP, 2012).

10.5 SOCIAL BENEFITS FROM NARMADA CANAL IRRIGATION

10.5.1 Positive Externalities of Narmada Irrigation

10.5.1.1 Energy-Saving Benefits for Well Irrigators in Command Area

Surveyed sample wells in the command area show, on average, the depth to water level fell across seasons (Table 10.2). The results suggest that after water from the Narmada project was supplied to those areas in 2002, there was a significant and consistent rise in water levels across seasons in all the six districts. This change might have occurred due to (1) decreased groundwater draft, resulting from reduced dependence on wells, and (2) groundwater recharge from return flows of gravity irrigation using canal water. In addition, the high rainfall received in most of these areas during the period from 2001 to 2006 might have influenced water level trends, though the effect of rainfall would be much lower as compared to imported surface water for gravity irrigation.

However, groundwater use in this tribal area of south Gujarat actually increased over time. More and more farmers belonging to the scheduled caste and scheduled tribes started investing in wells under government schemes in areas not receiving canal water from Narmada. This might have influenced the groundwater table in the region, which is probably the reason for the lower rise in water levels in command areas in the south Gujarat districts of Bharuch, Panchmahals, and Narmada as compared to north Gujarat districts.

Every 1 m of reduction in pumping depth reduces the energy use for pumping one cubic meter of groundwater by nearly 0.055 kWh. Based on this conversion ratio, the reduction in pumping depths($R_{P-DEPTH}$), the average volume of irrigation water applied per ha of well-irrigated area

Table 10.3 Average groundwater use for crop production in well command area and energy saving per ha of irrigation (m³/ha)

District	Volume of groundwater use per ha of cropped area post-Narmada	Average reduction in pumping depth (m)	Energy savings per m³ of groundwater pumped (kWh (C×0.055))	Total energy saving per ha (kWh)	Total economic benefit per ha (INR)
Ahmedabad	866.0	8.5	0.5	406.8	2034.0
Bharuch	1247.8	2.2	0.1	153.5	767.5
Mehsana	2542.6	13.1	0.7	1834.0	9170.0
Narmada	2665.1	3.8	0.2	551.4	2757.0
Panchmahals	1383.3	5.1	0.3	386.8	1934.0
Vadodara	3610.0	5.9	0.3	1170.7	5853.5

Source: Authors' own analysis based on primary data; sample size $N=291$.

($V_{GROUND-WATER}$); and the net well-irrigated area in Narmada canal command in ha ($A_{WELL-IRRIGATION}$), the total energy saving can be estimated as:

$$R_{P-DEPTH} \times 0.055 \times V_{GROUND-WATER} \times A_{WELL-IRRIGATION} \quad [10.1]$$

Estimates of the average volume of groundwater pumped by the farmers in each region and the energy saved per ha of well-irrigated area are provided in Table 10.3.

As Table 10.3 indicates, Mehsana had the highest energy savings per unit volume of groundwater pumped. The reason is that the reduction in pumping depth is highest in Mehsana. The volume of groundwater used per ha of land is also one of the highest at 2542 m³. Mehsana also had the highest total energy savings per ha of well-irrigated area at 1834 kWh. The second highest energy savings per ha is in Vadodara (1170.7 kWh). Vadodara also has the highest volume of groundwater use per ha of well-irrigated area (3610 m³), in spite of being less arid than Mehsana. It should be kept in mind that groundwater use is not only a function of climatic conditions, with arid climates demanding more water, but also a function of the cropping patterns selected by the farmers. While in areas like Ahmedabad and Mehsana, aridity would demand more water for agricultural produce, the cropping pattern is such that the overall water requirement is much less compared to south Gujarat, where water-intensive crops, such as wheat and paddy, are grown.

The economic equivalent of this energy saving benefit is equal to:

$$ECOBEN_{SAVE-ENERGY} = \propto \times A_{WELL-IRRIGATION} \times COST_{ENERGY} \quad [10.2]$$

Table 10.4 Changes in incidence of well failures and well command area in the Narmada command area after the introduction of canal irrigation

	Yearly incidence of well failures	
District	**Pre-Narmada**	**Post-Narmada**
Ahmedabad	3.0	1.2
Bharuch	4.8	5.2
Mehsana	0.0	1.2
Narmada	2.4	1.5
Panchmahals	0.0	4.0
Vadodara	9.5	0.0

Source: Authors' own analysis based on primary data collected from sample farmers in the six districts; sample size $N=291$.

where, $\propto$ is the total energy saved per annum per ha of well irrigation in kilowatt hour (column 5 of Table 10.3), and $COST_{ENERGY}$ is the cost of producing and supplying energy. This includes the direct economic cost of producing and supplying electricity and the cost of reducing carbon emissions from power generation, which is equal to INR[2] 0.47 per kilowatt hour of electricity. The value of $ECOBEN_{SAVE-ENERGY}$ can be obtained by multiplying the values in the last column of Table 10.3, by the area irrigated by wells in SSP command area in each district.

10.5.1.2 Reduction in Failure of Wells in the Command Area

In post-Narmada period, the incidence of well failures was reduced in four out of the six districts surveyed. This was found to be remarkable in Vadodara district, from 9.5 per year to almost nil. In Ahmedabad and Narmada districts, the reduction was lower. Contrary to this, in Panchmahals, there has been an increase in well failures from almost zero to 4.0 (Table 10.4). On the other hand, post-Narmada canal water, the area commanded by wells increased in four districts (Table 10.5) and remained the same in Bharuch. The negative trend in Panchmahals vis-à-vis the incidence of well failures and well command area could be due to the substantial increase in number of wells. This is perhaps disproportionately higher than the additional recharge available from the small canal-irrigated area, which results in well interference. Except for Mehsana, it can also be seen that, wherever the incidence of well failures decreased, the command area of wells increased, as expected.

10.5.1.3 Incremental Income of Well Irrigators From Crop Production

Positive externalities of canal irrigation as incremental income from well irrigation could be the result of: (1) an increased area under well irrigation

[2] One USD is equal to about Indian National Rupee (INR) 65.

Table 10.5 Well command area pre- and post-Narmada water introduction

	Well command area (ha)	
District name	**Pre-Narmada**	**Post-Narmada**
Ahmedabad	36.6	37.8
Bharuch	6.1	6.1
Mehsana	14.7	15.1
Narmada	2.4	3.1
Panchmahals	5.1	3.9
Vadodara	4.4	4.5

Source: Authors' own analysis based on data from field survey.

due to improved well yields; (2) shifts in cropping patterns toward water-intensive, high-value crops; (3) improved irrigation leading to higher yields; and (4) reductions in costs of irrigation due to decreases in pumping depths and lower costs of well deepening. Another important source of income for the well-owning farmers comes from higher water sales.

These findings reflect an assessment of whether there has been any change in cultivated area for various crops grown by the well-owning farmers who are the beneficiaries. The analysis of estimated average area under different crops of the sample well-owning farmers, covering all the three seasons, pre- and post-Narmada canal water, showed reduction in area more for rain-fed crops, namely, green gram, chick pea, maize, and paddy. The area expansion is more for irrigated crops as cotton and castor, and those sown in winter such as wheat in some locations, and water-intensive sugarcane in Bharuch.

Synthesizing the data presented in Table 10.5 and the observed changes in cropped area of well irrigators, it appears that the trend vis-à-vis cropped area of individual farmers is not in conformity with the trends noticed vis-à-vis the command area and water-level trends, particularly for Ahmedabad and Mehsana. While the water levels improved, and well command area increased, the gross cropped area has reduced slightly. This is perhaps because farmers do not own wells individually in such areas owing to very high capital cost of well construction. In these areas, over a period of time, the number of shareholders of tube wells increased, while the area irrigated by individual farmers reduced. Reduction in average land-holding size is also another reason for a reduction in the size of gross cropped area. Another possibility is that with greater reliability of groundwater resources, farmers replace the rain-fed crops by irrigated ones, but grow in smaller areas. Whether this is the correct trend is examined in the later part of the section.

There was a significant increase in area for irrigated cotton and castor in Panchmahals, while the area under paddy cultivation, which is mostly rain-fed, fell. In Panchmahals the formations underlying the command areas are consolidated rocks (Gujarat Ecology Commission, 1997). In these consolidated formations, the aquifer storage potential is generally very poor, and therefore, improvement in groundwater recharge through irrigation return flows from canals would bring about a marked difference in groundwater tables, making well irrigation more sustainable. Yet, this benefit of improved recharge is not reflected in the command area, perhaps because the farmers are irrigating their crops more intensively or there is an overall increase in the number of wells in the area. The second highest increase in irrigated areas was observed in Vadodara, with an increase in irrigated area of wheat (0.3 ha) and castor (0.12 ha).

Nevertheless, this change in cropping pattern is a result of improved availability of water in the wells. Analysis of crop yields in different seasons in the well command shows crops grown in the well commands of the six locations had yielded changes, mostly positive. Only winter maize (in one location) and green gram (in another location) showed some yield reductions. The only two crops that did not show any changes in yields were groundnut and summer maize. For all other crops, changes were recorded at least in one district, and in some cases, in more than four crops. The percentage increase in yield was very high (32.8) for castor in Bharuch. The increase in cotton yields was the lowest, at 27%, for Mehsana and the highest, at 60.9%, for the Narmada district. In the case of paddy, the highest yield increase of 90.9% was found in Panchmahals. This must have been the result of increased input use, as supported by the empirical data from the field. For instance, the intensity of groundwater use (m^3/ha of cropped land) in the well command areas had increased post-Narmada, clearly indicating that either farmers were applying more water to their irrigated winter crops or they were applying supplementary irrigation to crops otherwise grown under rain-fed conditions (Table 10.6). These suggestions could have an impact on yields.

To know how this has impacted the overall net returns from these crops, the changes in cost were analyzed. The analysis shows that the cost of cultivation has increased for all the crops except for tobacco in Narmada district, in which case the yield did not show any improvements. The cost of cultivation includes the cost of all inputs, namely, plowing, sowing, seeds, fertilizers, and pesticides.

Table 10.6 Average groundwater use per ha of cropped area in well command (pre- and post-Narmada)

	Average groundwater use (m³/ha)	
District name	**Pre-Narmada**	**Post-Narmada**
Ahmedabad	867.90	865.99
Bharuch	821.76	1247.78
Mehsana	2491.84	2542.61
Narmada	1689.26	2665.14
Panchmahals	1201.31	1383.34
Vadodara	2490.94	3609.88

Source: Authors' own analysis based on data from field survey.

Table 10.7 shows the changes in net income from the crops. It shows that for all the crops, the farmers' net income from crop production has increased. The highest income increase was for cotton, with values ranging from INR 42,212/ha (US$ 700/ha) in Bharuch to INR 71,030/ha (US$ 1180/ha) in Vadodara. This is followed by castor, for which the highest increase in net income was of INR 50,734/ha, in the Mehsana district, famous for oil seed production. In the case of green gram, a slight reduction in income per ha was found. It is important to remember that this crop generally grows under residual soil moisture and does not require many agronomic inputs as a result of which yields and returns would not change significantly as a result of improved well water availability. In the case of summer bajra, a major reduction in net income was identified as no yield improvements were recorded for this crop and average inputs costs went up.

An astonishingly high positive net income change of INR 806,391/ha was recorded for tomato in Mehsana. However, such values are unrealistic as farmers grow such crops in very small areas and a significant increase in its volatile market price can raise returns from a cultivated hectare. Nevertheless, the increase in net income is quite substantial for most crops, across districts.

10.5.1.4 Incremental Income of Well Irrigators From Dairy Production

Dairy production is a very important activity in rural Gujarat, and intensive dairy farming is practiced in most districts currently covered by the SSP (Singh et al., 2004). Improved groundwater recharge, without a proportional increase in the number of wells, may result in an effective increase in average command area of individual wells. This in turn can potentially

Table 10.7 Change in net income of well-owning farmers from crop production

Season	Crop	District change in net income from crop production in well-irrigated command (INR/ha)					
		Ahmedabad	Bharuch	Mehsana	Narmada	Panchmahals	Vadodara
Kharif	Paddy	23,467.4		−1181.9	925.3	12,351.6	12,675.2
	Cotton		42,212.0	67,733.9	48,289.0	65,293.9	71,029.9
	Chick pea		17,871.3		14,152.8	11,103.3	16,539.4
	Jowar		7449.4				
	Green gram		15,187.7	−2847.0			
	Castor		102,755.0	50,734.6		35,521.8	35,553.0
	Sugarcane						
	Maize			21,900.0	11,657.7	5310.7	12,599.5
	Bajra			14,085.3		10,087.1	
	Tomato			187,851.0			
	Tobacco						
	Banana					29,448.2	
Winter	Wheat			17,285.5			41,587.6
	Cumin	4408.3	6478.1	49,316.4	5006.8	5851.0	4506.0
	Fennel			54,708.4			
	Tomato			806,391.0			
	Maize						
Summer	Jowar				6727.6	11,169.0	52,131.4
	Bajra	−13,906.5					
	Maize						

Source: Authors' own analysis based on data collected through a primary survey.

influence the animal husbandry activities of the well owners as there would be more biomass available in the farm, in the form of dry and green fodder, and more water for livestock drinking. Our analysis shows that the holding of indigenous cows increased in Ahmedabad, Mehsana, and Narmada, and marginally in Panchmahals and Vadodara, whereas it reduced in Bharuch. As regards crossbred cows, the holding size increased significantly in Vadodara, and marginally from nil in Bharuch. A reduction (from 0.11 to 0.05) was seen in Ahmedabad. As regards buffalo, its population size increased in Ahmedabad, Mehsana, Vadodara, and Narmada, while it reduced in Bharuch and Panchmahals. Overall, the trend in terms of livestock holding appears to be positive in most locations, except Bharuch. This is consistent with the trend in cropped area, wherein the gross cropped area under well command increased everywhere except Bharuch.

The absence of any change in composition of livestock held by farmers in Bharuch needs further explanation. In the case of Bharuch, the biggest change in cropping system is in terms of shift from short-duration, rain-fed varieties of cotton and castor to long-duration irrigated varieties and supplementary irrigation of crops, namely, jowar, chick pea, and green gram. Along with these, there has been a reduction in area of all these crops. Though the yield of these crops had improved, the overall output may not have increased. That said, the positive externality of canal water on well irrigation cannot be assessed in relation to what happens for an individual well command as the number of wells has increased over the years.

The average income of dairy farmers has also changed significantly for all livestock types in all locations, except for indigenous cows in Bharuch and Mehsana (Table 10.8). The positive trend vis-à-vis buffaloes and crossbred cows can be attributed to increased production of crops which yield dry or green leafy biomass, such as paddy, jowar, wheat, bajra, and green gram, which are used as fodder[1]; and sometimes those which have byproducts that can be used as feed for animals such as green gram and groundnut. Another important factor that might have contributed to the increased yield of dairy animals is the better care they might be getting with a general improvement in the welfare of the farmers keeping these animals. But, the reduction in net income in the case of indigenous cows may be because of less interest farmers probably show in rearing low-yielding indigenous varieties, which are more hardy and preferred in situations of water and biomass shortages.

Table 10.8 Average income of well-owning farmers from dairy production (INR/animal/lactation) before and after water supplies from the Narmada canal

	District					
Type of animal	**Ahmadabad**	**Bharuch**	**Mehsana**	**Narmada**	**Panchmahals**	**Vadodara**
			Before Narmada			
Indigenous cow	5178.74	5420.00	13,187.50	4680.00	0.00	2002.50
Crossbred cow	8255.38	0.00	0.00	0.00	0.00	11,740.00
Buffalo	13,473.57	2000.00	15,017.60	8750.37	5018.38	5941.14
			After Narmada			
Indigenous cow	15,859.38	5215.00	12,805.90	11,950.00	13,080.00	7438.13
Crossbred cow	14,567.34	0.00	0.00	0.00	0.00	18,340.00
Buffalo	24,921.93	12,475.00	24,571.40	13,722.50	12,033.70	15,255.00
			Change in income			
Indigenous cow	10,680.64	−205.00	−381.60	7270.00	13,080.00	5435.63
Crossbred cow	6311.96	0.00	0.00	0.00	0.00	6600.00
Buffalo	11,448.36	10,475.00	9553.80	4972.13	7015.32	9313.86

Source: Authors' own analysis based on primary data.

10.5.1.5 Incremental Farm Surplus of Well Irrigators due to Canal Irrigation

The economic value of positive externality of canal irrigation on groundwater can be estimated using the following output variables.

The current gross groundwater-irrigated area in the command area of canals and areas receiving irrigation through canal lift (A_{WELL}); the average increase in net income from well irrigation per unit of the gross well-irrigated area $\left(\varnothing^{1}_{WELL}\right)$; increase in groundwater-irrigated area post–canal water introduction (ΔA_{WELL}); net return from groundwater irrigation pre-canal water introduction ($\varnothing_{WELL}$); and the current average livestock population per ha of gross well-irrigated area (N_{WELL}); and the increase in average net income from livestock production of well irrigators per unit of livestock for all types of livestock owned by the well owners $\left(\beta^{1}_{WELL}\right)$; and average increase in livestock population per ha of gross well-irrigated area $\left(N^{1}_{WELL}\right)$ as:

$$WELL_{EXTERN} = A_{WELL} \times \varnothing^{1}_{WELL} + \Delta A_{WELL} \times \varnothing_{WELL} + N_{WELL} \times A_{WELL} \times \beta^{1}_{WELL} + \Delta N_{WELL} \times A_{WELL} \times \beta_{WELL} \quad [10.3]$$

The same for a unit of gross well-irrigated area can be expressed as:

$$WELL_{EXTERN-UNIT} = \varnothing^{1}_{WELL} + \{\Delta A_{WELL}/A_{WELL}\} \times \varnothing_{WELL} + N_{WELL} \times \beta^{1}_{WELL} + \Delta N_{WELL} \times \beta_{WELL} \quad [10.4]$$

The estimates of incremental farm surplus of well irrigators did not include Surendranagar, as the region's groundwater is totally saline, with that irrigation from Narmada is not expected to make any positive impact on the region's saline aquifers in the medium term. This is the positive externality of canal irrigation on well irrigation. The value (in terms of incremental income surplus for well irrigators per ha of gross well-irrigated area) was found to be highest in Ahmedabad (INR 113,587), and lowest in Panchmahals (INR 24,719). The incremental surplus per ha of well irrigation was INR 46,641 for Mehsana, INR 47,422 for Vadodara, and INR 49,176 for Narmada.

Analysis shows that while the average area under well irrigation declined in four out of the six cases, the average net income from crop production substantially increased, with the average incremental income often exceeding the average net income prior to Narmada waters. Here, it must be mentioned that the average net income is the weighted average of the net income from various crops, estimated on the basis of the area under each of the crops. Therefore, the shift in cropping pattern toward high-value irrigated crops has definitely influenced the overall net income from unit area under crops. On the other hand, dairying can also influence farm surplus.

Marginal increase in the number of livestock is noticed along with a substantial increase in average net income from dairy production per unit of livestock kept by the well irrigators, post-Narmada water introduction.

It is theoretically incorrect to translate improved groundwater balance into increased irrigation potential of wells and actual well-irrigated area in the canal command, when it is merely because of reduced draft, and not because of increased recharge from return flows. The reason is the utilizable groundwater resources are not enhanced there. Therefore, reduction in the well-irrigated area found in our survey is quite natural in at least some of the locations.

10.5.1.6 Reduction in Cost of Domestic Water Supplies in the Command Area Villages and Towns

The pipeline network based on Sardar Sarovar Narmada canal and Mahi canal is expected to serve nearly 54% of the villages in Gujarat, apart from 125 towns (source: based on www.gwssb.org/pdf/narmadaprojects.pdf.). The entire Saurashtra and Kachchh and most parts of north and central Gujarat are to be covered under this scheme. But, there are thousands of villages in south and east Gujarat which are not to be covered by this pipeline network. These areas are generally being considered as the well-endowed region of the state in terms of water resources.

However, even within south Gujarat, there are two distinct geohydrological environments. The first falls in the alluvial belt covering parts of Vadodara, Narmada, Surat, and Bharuch. This area does not face problems of physical shortage of water during any part of the year, owing to the rich alluvial aquifers in the area which get sufficient replenishment from rainfall, and plenty of surface water in ponds, though there are problems of water quality. The other falls in the hard rock region, with consolidated rocks of Deccan trap basalt. Parts of Vadodara, Narmada, and Bharuch, which fall inside the Narmada command, are in the Deccan trap area. Panchmahals in east Gujarat is fully underlain by crystalline hard rock formations. In spite of moderate to high rainfalls, this hard rock region has poor groundwater potential (Gujarat Ecology Commission, 1997). Drinking water shortage due to drying up of wells occurs in summer months.

Like irrigation wells, the drinking water supply sources located inside the command area and which tap local aquifers are likely to be benefited by canal irrigation through return flows from irrigated fields and seepage from canals. These benefits are affected in the following way: (1) rise in water levels in the wells, which can reduce the cost of pumping groundwater, thereby reducing the operational cost of village Panchayats running these schemes; (2) reduced incidence of well failures and therefore reduced investment for well deepening, etc.; (3) rejuvenation and improved yield of wells, resulting in increase in level of supply of water, especially during summer months; and (4) dilution of the minerals present in the groundwater, thereby improving the quality of drinking water supplied by the local well-based schemes. While the benefits discussed under items "1" and "2" can be put into monetary terms, evaluating the third and fourth types of benefits would require estimation of water demand functions and this method is attempted here.

The amount of money required to be spent to transport water from a distant source to meet the water supply needs during summer months in the face of drinking water shortage can be considered as the economic benefit from improved yield of drinking water wells and sustainable drinking water supply. Furthermore, the economic benefits from improved chemical quality of water in the drinking water wells can be treated as the same as the cost of treating the poor-quality water, through treatment processes like desalination or defluoridation, depending on the case. While the first three types of impacts of canal water introduction would be more visible in hard rock regions, because of the unique nature of aquifers in those regions, the fourth type of impact, ie, dilution of minerals present in groundwater, is more likely to be seen in the alluvial areas of north Gujarat. However, all the villages surveyed for canal lift in Ahmedabad and Mehsana were found to have been covered by the Narmada canal–based pipeline scheme, and therefore such impacts on groundwater quality may not translate into the type of positive externalities discussed above. Therefore, the social benefits of canal irrigation in the form of improved sustainability of groundwater-based drinking water schemes would be affected in the south and east Gujarat parts of Narmada command.

The economic benefit from this externality can be computed as:

$$P_{WELL}\,(R_{P\text{-}DEPTH} \times 0.055 \times V_{DRINK\text{-}WATER} \times COST_{ELECT} + V_{DRINK\text{-}SUMMER} \times PRICE_{TANKER}) \quad [10.5]$$

Here, $R_{P-DEPTH}$ is as defined earlier for Eq. [10.2]. $V_{DRINK-WATER}$ is the per capita volume of water used for drinking in a year in m^3; P_{WELL} is the total size of the population dependent on groundwater-based schemes for domestic water supply in the region; $COST_{ENERGY}$ is the average cost of generating and supplying 1 kW of electricity; $V_{DRINK-SUMMER}$ is the per capita volume of water used for drinking and domestic uses in summer; and $PRICE_{TANKER}$ is the market price of a cubic meter of tanker water supplied.

The reduction in pumping depth is estimated to be 12.5 ft, ie, 3.75 m (source: based on Table 10.3). The cost of generating and supplying energy for groundwater pumping is taken as INR 5 per kilowatt hour. The volume of water required to meet the domestic water supply needs is estimated to be 36.5 m^3 per capita for the whole year[3]. While it has to be met from wells during most parts of the year, during summer months, only part of this demand will be met from wells, and water will have to be supplied though tankers at the rate of 40 lpcd in the absence of surface irrigation from Narmada canals. For tanker water, the on-going market price of INR 50 per m^3 was considered. As a result, water delivery through Narmada canals would augment the water supplies in wells so as to avoid the dependence on tankers. Furthermore, the population depending on groundwater-based drinking water supply sources, in the scarcity-hit region, at present was considered as 5 million. Thus, the total economic benefit is estimated to be 1069.4 million rupees per annum ($213.93 \times 5 \times 10^6 = 1093.3$ m rupees).

In addition to the benefits from reduction in pumping depths and improvement in water supply quantity, benefits also accrued from improvement in quality of water, occurring due to dilution of groundwater. A survey carried out in one of the villages in Vadodara district, which receives canal water for irrigation, shows significant improvement in the quality of groundwater used for domestic purpose. The area is underlain by basalt formations. The community in the village is dependent on bore wells drilled on the banks of Orson River for domestic water supply. There is a notable reduction in the number of families reporting problems of water quality in the sources and they attribute the quality improvement to dilution of minerals present in groundwater across the seasons. No household reported high TDS in water used for drinking and cooking in their public water sources, which include the hand pumps in the locality, after the introduction of Narmada canal water in the village, while at least five households encountered problems with their drinking water source, ie, the distant bore well located on the river bank prior to the introduction of Narmada waters[4].

But the village households use water from local bore wells and hand pumps, which is available through individual tap connections and stand posts, for domestic water supply (other than drinking and cooking) and livestock drinking. Twenty-five out of the 30 families surveyed reported problems with their domestic water sources during winter and summer

[3] This is based on the assumption that the communities would require 100 L of water per day.

[4] The low incidence of poor water quality is because the bore well which was tapped for drinking water supply used to be replenished continuously by the river water, which is free from salts.

months prior to the introduction of Narmada canal water in the village, with no families complaining about water quality during the rainy season. However, no families reported the problems of salinity in water after Narmada canal water was introduced. This is quite possibly because of the fact that the wells get recharged during monsoon with the freshwater remaining on the top of the water table. This water gets used up during the rainy season, with the water having salinity remaining in the lower strata of the consolidated formations and this water gets used up during winter and summer months. After the introduction of Narmada canal, the wells are continuously recharged by the return flows from canal irrigation.

10.6 FINDINGS AND CONCLUSIONS

The supply of surface water from the Narmada canal for irrigation has resulted in changes in groundwater level trends over time. Similarly, there has been a marked reduction in the salinity of groundwater over time. Irrigation wells located in the command area of the Narmada canal are found to have benefited from seepage. This is manifested by a consistent rise in groundwater levels across seasons, resulting in reduced depth of pumping. The highest average rise in well water level during monsoon (47 ft) and winter (49 ft) was found in wells located in Mehsana, and during summer (53 ft) in Ahmedabad. There has been a notable reduction in incidence of well failures in the canal command area and an increase in command areas of wells post-Narmada.

The reduction in well failure was found to be very remarkable in Vadodara district, from 9.5 per year to almost nil today. But, the same trend was not visible in Panchmahals district. A probable reason can be the substantial increase in the number of wells tapping the hard rock aquifers of the district over the past few years, which perhaps is disproportionately higher than the additional recharge available from the small canal-irrigated area, resulting in well interference. These changes produce multiple water-use benefits, in the form of irrigated crop production, dairy farming, and domestic water supplies.

First of all, the cropping pattern of well irrigators in the command had changed; there was substantial increase in yield and income returns per unit of land (INR/ha) for well irrigators. While there were incidences of major reduction in the area for the "normally" rain-fed crops of the kharif season, the area under irrigated crops such as cotton, castor, and wheat had increased in some locations, and the same result is visible for water-intensive crops such as sugarcane in Bharuch. Almost every crop grown in the well command areas of the six districts showed mostly positive yield changes after the introduction of Narmada waters, which could be explained by the greater dosage of irrigation to these crops, as reflected in the increase in the average volume of groundwater applied by the farmers per ha of land post-Narmada. There was a marginal increase in the livestock-holding for the well irrigators post-Narmada water. Overall, the trend in terms of livestock-holding appears to be positive in most locations, except Bharuch.

The indirect benefit of reduced economic cost of energy used for well irrigation was huge. For every hectare of well-irrigated area in SSP command, the benefit to the society through energy saving ranges from INR 768 to INR 9170 in Mehsana. The indirect economic impact of canal irrigation in the form of improved sustainability of groundwater-based drinking water supply schemes in south and east Gujarat parts of Narmada command for a population of five million is estimated to be INR 1032.5 million rupees per annum. The primary survey conducted in Vadodara district showed improvement in the quality of groundwater used for domestic water supply.

In the past, many large- and medium-sized water resources projects in India had been criticized for their poor performance in terms of area irrigated. However, too little attention has been paid to the multiple water-use benefits. Our analysis of the SSP shows that the multiple-use benefits from the large water system, by improving drinking water supply, dairy production, and well irrigation are extremely important.

An important benefit that can be derived from the water supplied from the Sardar Sarovar reservoir is inland fish production. The reservoir fishery is already happening in Sardar Sarovar reservoir, through fishery cooperatives of the local fishing community. However, its productivity would be limited. Large reservoir fishery is facing problems of low productivity in India, due to rapid drawdown in water levels experienced after the monsoon, when the water level would be highest (Jhingram, 1988). In order to raise productivity levels in large reservoirs, fisheries scientists and managers need to understand the average pattern of irrigation water demand so as to adjust their stocking and harvesting programs accordingly (Saha and Paul, 2000). An alternative is to store a small share of the water released from the reservoir during monsoon and winter season in shallow ponds in the command area, which can act as reservoirs for freshwater fisheries and can be stocked with fingerlings. In many areas such as Mahi command in south Gujarat, Hirakud command in Odisha and in the deltas of Krishna and Godavari, the water from canals is used to replenish local ponds for raising fish.

REFERENCES

Alagh, Y.K., 2010. A Sardar Sarovar Riddle. Financial Express, Ahmedabad, Gujarat.

Allen, R.G., Willardson, L.S., Frederiksen, H., 1998. Water use definitions and their use for assessing the impacts of water conservation. In: Proceedings of ICID Workshop on Sustainable Irrigation in Areas of Water Scarcity and Drought. England, Oxford, pp. 72–82.

Biswas, A.K., 1994. Sustainable water resources development: some personal thoughts. International Journal of Water Resources Development 10 (2), 109–116.

Biswas, A.K., Tortajada, C., 2001. Development and large dams: a global Perspective. International Journal of Water Resources Development 17 (1), 9–21.

Biswas-Tortajada, A., 2014. Gujarat water supply grid: step towards achieving water security. International Journal of Water Resources Development 30 (1), 78–90.

Chakravorty, U., Umetsu, C., 2003. Basin-wide water management: a spatial model. Journal of Environmental Economics and Management 45 (1), 1–23.

Daines, S.R., Pawar, J.R., 1987. Economic Returns to Irrigation in India. SDR Research Groups Inc. & Development Group Inc, New Delhi.

Dhawan, B.D., 2000. Drip Irrigation: Evaluating Returns, Economic and Political Weekly 3775–3780 October 14.

Gujarat Ecology Commission, 1997. Eco Regions of Gujarat. Gujarat Ecology Commission, Vadodara.

Hernandez-Mora, N., Llamas, R., Martinez-Cortina, L., 1999. Misconceptions in aquifer overexploitation implications for water policy in southern Europe. In: Agricultural Use of Groundwater. Springer, The Netherlands, pp. 107–126.

Institute for Resource Analysis and Policy (IRAP), 2012. Realistic Versus Mechanistic: Analysing the Real Economic and Social Benefits of Sardar Sarovar Narmada Project Report submitted to SSNNL. Institute for Resource Analysis and Policy, Gandhinagar, Hyderabad.

Institute for Resource Analysis and Policy (IRAP), 2015. Water Resources Development, Use and their Management in Andhra Pradesh-A study (prepared for the Agricultural Commission of Andhra Pradesh), draft submitted to Centre for Economic and Social Studies (CESS), Institute for Resource Analysis and Policy, Hyderabad-82.

Institute of Rural Management Anand / UNICEF, 2001. White paper on Water in Gujarat, Final Report Submitted to the Dept. of Narmada, Water Resources and Water Supply, Government of Gujarat, Gandhinagar, Institute of Rural Management, Anand.

Jhingran, A.G., 1988. Reservoir Fisheries Management in India. Bulletin 45. CICFRI, Barrackpur, India.

Kumar, M.D., 2007. Groundwater Management in India: Physical, Institutional and Policy Alternatives. Sage Publications, New Delhi.

Kumar, M.D., 2009. Water Management in India: What Works, What Doesn't. Gyan Books, New Delhi.

Kumar, M.D., van Dam, Jos, C., 2013. Drivers of Change in Agricultural Water Productivity and its Improvement at Basin Scale in Developing Economies. Water International 38 (3), 312–325.

Meinzen-Dick, R.S., 1995. Timeliness of irrigation: performance indicators and impact on agricultural production in the Sone Irrigation System, Bihar. Irrigation and Drainage Systems 9 (4), 371–387.

Molle, F., Turral, H., 2004. Demand management in a basin perspective: is the potential for water saving overestimated? In: Paper Presented at the International Water Demand Management Conference, Dead Sea, Jordan.

Parasuraman, S., Upadhyaya, H., Balasubramanian, G., 2010. Sardar Sarovar project: the war of attrition. Economic & Political Weekly 45 (5), 39–48.

Ranade, R., Kumar, M.D., 2004. Narmada water for groundwater recharge in North Gujarat: conjunctive management in large irrigation projects. Economic and Political Weekly 39 (31), 3510–3513.

Saha, C., Paul, B.N., 2000. Flow through system for industrial aquaculture in India. Aquaculture Asia 5 (4), 24–26.

Seckler, D., 1996. The New Era of Water Resources Management: From "Dry" to "Wet" Water Savings Research Report 1. International Irrigation Management Institute, Colombo, Sri Lanka.

Shah, Z., Kumar, M.D., 2008. In the midst of the large dam controversy: objectives, criteria for assessing large water storages in the developing world. Water Resources Management 22 (12), 1799–1824.

Shah, T., Gulati, A., Hemant, P., Shreedhar, G., Jain, R.C., 2009. Secret of Gujarat's agrarian miracle after 2000. Economic and Political Weekly 44 (52), 45–55.

Singh, O.P., Sharma, A., Singh, R., Shah, T., 2004. Virtual water trade in dairy economy: irrigation water productivity in Gujarat. Economic and Political Weekly 39 (31), 3492–3497.

Talati, J., Shah, T., 2004. Institutional vacuum in Sardar Sarovar project: framing 'rules-of-the-game. Economic and Political Weekly 39 (31), 3504–3509.

Verghese, B.G., 2001. Sardar Sarovar project revalidated by supreme court. International Journal of Water Resources Development 17 (1), 79–88.

Watt, J., 2008. The Effect of Irrigation on Surface-ground Water Interactions: Quantifying Time Dependent Spatial Dynamics in Irrigation Systems Unpublished Doctoral Thesis. School of Environmental Sciences, Faculty of Sciences, Charles Sturt University, Sydney, NSW, Australia.

CHAPTER 11

Re-Imagining the Future: Experiencing Sustained Drinking Water for All

A.J. James
Institute of Development Studies, Jaipur, India

Reproduced with permission from the India Today issue of 12 November 2025[1]

11.1 INTRODUCTION

Although I have just completed my first field assignment in India, after more than a decade of living and working abroad, I am yet to fully comprehend the scale and depth of the transformation that some visionary and smart governance has done to the rural water supply situation in the country. Refusing to believe the glowing media accounts I had read, I was determined to see for myself and "uncover the truth," which had been the hallmark of my earlier work in India. But in village after village (I visited a total of 100 Gram Panchayats across 10 states in 6 months); a truly astonishing picture was emerging. Villagers had enough water for domestic use (drinking, cooking, cleaning, and bathing), their livestock had water, and emergency water supply in trains and tankers during summer scarcities and droughts were a thing of the past. The community-managed systems were supported by government departments working in close coordination with each other and this combination has helped them to address the various problems encountered en route to their common goal. While this had been unheard of in the earlier India I knew, where only a few "islands of excellence" survived, the reasons for the success were in the details. And it was these details that the other developing countries in Africa wanted, in order to learn from India's remarkable success in overcoming a decades-old problem. This in brief was my

[1] This is a fictional account, written as a journalistic piece published in a popular magazine, but supported wherever relevant by available reports and documents, which are mentioned in the footnotes.

Rural Water Systems for Multiple Uses and Livelihood Security
ISBN 978-0-12-804132-1
http://dx.doi.org/10.1016/B978-0-12-804132-1.00011-1

assignment, and here are the highlights, extracted from the voluminous report that was submitted to the funding agency.

11.2 VILLAGE WATER MANAGEMENT: LOCAL INGENUITY MEETS SUPPORTIVE GOVERNMENT

Forty-year-old Gopal Rao of Brahmasamudram village in Andhra Pradesh described the process as he knew it. "Some government people came along with a local NGO and asked us to help them manage our tanks. We were reluctant, as we had forgotten the old ways, and the young people in the village had migrated to urban areas in search of jobs. The tank in our village had been dry for several years and the high cost of labour meant that most of our fields were lying uncultivated. But the government people and the NGO staff were persistent. They talked to the elders, who remembered the tank being full and like a sea—which is how our village got its name, and persuaded us to try.[2] We had several meetings to plan how to revive the inflow channels, renovate the tank sluices and bunds and clean the channel that led to the tank in the downstream village of Anantasagaram. We made a plan, and the government officers checked it and told us to make changes, and after several weeks, it was finally ready. Then there was a big meeting where officials from different departments came and agreed how the work should be done. Some of the old check dams were broken, and for others we made gates, so that the initial monsoon water could flow down to the tank.[3] We worked for six months and finally it was ready. That year the monsoon was good, and the tank filled again."[4]

The prosperity in this and other villages was evident from a transect walk. Although it was the winter season (October–March), almost all fields were under cultivation, with drip and sprinkler systems in almost all of them, and evidenced the growth in Rabi cultivation reported in the

[2] The word 'samudram' in the name Brahmasamudram and the word 'sagaram' in the name Anantasagaram, both of which are names of actual villages in Andhra Pradesh, means sea or ocean - referring to the size of the tanks that had been built there which, when full, looked like a sea.

[3] The idea of gated check dams was already integrated into the design of the famous Kohlapur Type (KT) Weir (see Michael, 2009, p. 69) but not in smaller check dams with a catchment of less than 1 hectare in area. The idea of putting gates into these smaller check dams was first discussed in WHiRL (2004) and subsequently finds mention in other government guidelines (e.g., GoO, 2010).

[4] This account is based on an experiment in Anantapur district in the early 2000s, where the then collector, Shri. Somesh Kumar IAS, had renovated a cascade of tanks, based on old descriptions, and that year the tanks had filled. According to him, an elderly villager fell at his feet saying he must be a demigod as he had last seen the tank full only as a young boy (Somesh Kumar, personal communication, 2003).

secondary statistics (from around 50% of total cultivated area on average in the state in 2015 to around 80%).[5] In addition to traditional crops like paddy and jowar, there were newer varieties (like Madagascar rice also called System of Rice Intensification or SRI) and new crops (such as millets and mustard), seed multiplication units and vegetables being grown in poly-houses, all of which used less water per unit of land than the earlier crops. Ground water consumption for irrigation was being closely monitored with piezometers put in by the Groundwater Department, and groundwater levels had actually been *rising* in the last 3 years, two of which brought less than average rainfall.

11.3 BUILDING SUPPORT FOR CHANGE

Janardhan, the local resource person of the implementing NGO, explained this phenomenon: "Around 100 years ago, even if the monsoon was poor, there was always water in the wells, since groundwater was not being used. But after the drilling of bore wells to tap ground-water, from the 1970s, the aquifers began to be used up very fast. When the advocacy by NGOs, academics and field workers failed to stop the over-extraction in this area, a group of us decided to do something about different: we formed a coalition of NGOs and resource persons – called ourselves the *Imagine India Coalition*, collected and analyzed information, formulated our arguments, and focused on the politicians. We made presentations in various Legislative Assemblies across the country, showed scientific data from our top scientists, took Members of the Legislative Assembly (MLAs) and Members of Parliament (MPs) on exposure visits to affected areas – and engaged them in public discussion on TV and radio and newspapers. When they understood that the water situation could easily improve with a few specific changes to the incentives facing farmers, the scenario changed. They pushed for changes in policies, programmes and laws – and that made all the difference!"

The situation (and process) was similar in most of the other villages in all the other states. This surprised me. In the past, one state had been reluctant to "learn" from another—often claiming that their state was "unique" and preferring to evolve their "own" solutions—no doubt, in an effort to garner the distinction of doing something new for the first time in their state, rather than the less personally rewarding option of adapting an effective

[5] The area under cultivation during the winter (Rabi) season in 2015 was around 50% of the total cultivated area.

solution from what had been tried elsewhere.[6] (I was to learn later (see below) that this was due to the strong leadership exercised by the central government, which had strongly supported the engagement with elected representatives.)

11.4 CO-MANAGING WATER RESOURCES

The local villagers had then sat down with the government officials and, with the active support of Janardhan and his team, chalked out their plan for conserving water resources. At first, they focused on water conservation and groundwater recharge. They worked out a plan to install rainwater-harvesting structures on the roof of the school and the Panchayat office, which channeled monsoon water to deep shafts in the school premises. Then they studied the flow of wastewater in the village and built channels to direct these into abandoned open wells in different parts of the village. But piezometer readings showed that the groundwater levels were still declining year on year! This was a big disappointment for them.[7]

In the next round of discussions, however, the villagers themselves analyzed the problem and finally agreed that over-extraction from agricultural bore wells was the main reason why groundwater levels were going down every year. This, Gopal Rao explained, was something that farmers had known for a long time, but did not want to mention for fear that it would lead to measures to curtail their water! But when pushed by Janardhan and his team on why groundwater levels were still falling, they had no option but to admit that this was indeed the real culprit.

Janardhan said that this led to a lot of discussions within the team and with the government officials locally and also in the district headquarters and the state capital. After nearly a year of listening to experts, reading many reports and articles on the issue, and visiting other areas, a policy had been formulated—based on two simple objectives:

6 This was especially the case with the Water Supply Management Organization (WASMO) of the Government of Gujarat, set up in 2002, which had achieved the unprecedented distinction of facilitating community management of domestic water supplies in around 16,000 out of 18,000 GPs in Gujarat by 2012. Although the central government had tried to replicate this phenomenon by creating generously funded Water Supply and Sanitation Organizations (WSSOs) in every state, no other state had neared the achievements of WASMO even by 2015. See James (2011) for a detailed analysis of WASMO and other models or rural drinking water supply schemes in India, and www.wasmo.org for details of its achievements.

7 This is the case of Ismailpur village (of Sopura Gram Panchayat in Mandawa block of Jhunjhunu district in Rajasthan), detailed in James et al. (2015b). Although the village had implemented impressive roof rain water-harvesting systems, diversion of water flowing across roads and open areas to wells for groundwater recharge, and a channel network to collect and send household wastewater to a collection tank for horticulture, piezometer readings showed a continued decline in groundwater levels in the village.

- Reduce crop water use, but maintain farmer profit;
- Reduce groundwater use in every village.

Government departments had been tasked with finding ways to achieve these two objectives.

However, real progress only came when the NGO coalition group (The *Imagine India Coalition*), with the help of the local media, focused their attention on getting the support of the MPs and MLAs of various state governments. After making presentations to this group of elected representatives, taking them on exposure visits and having one-on-one meetings with each MLA, they finally managed to convince them to take a public stand in support of a National Water Policy based on these two core objectives, which was subsequently endorsed in State Water Policies.[8] (A third core objective was co-management of water resources by community and government, with institutional support provided by the government to water resource planning and management by committees including elected representatives from Gram Panchayats and district-level Zilla Parishads.)

While the announcement of the new Water Policies in a record time of less than 1 year was a considerable achievement, their crowning achievement was in convincing various Chief Ministers and their cabinet of Ministers to support the new policies and future government programs to implement them. With MPs, MLAs, and the Ministers convinced, the bureaucrats and engineers in various departments found it much easier to work, and real progress began.

11.5 COMPETING TO REDUCE GROUNDWATER EXTRACTION

The key driver at the village level, according to Janardhan, was the incentive policy for reduced extraction from bore wells. The state-level "Water Warriors"[9] competition made it mandatory for each participating Gram Panchayat (GP) to show reduced piezometer readings, the prize going to villages with the best local arrangements for self-sufficiency in domestic

[8] James (2011) concluded that the key factors underlying the success of the Rural Water Supply Service Delivery Models in India are: "the motivation levels of senior bureaucrats and politicians in showing each project to be a success; the support provided by external funding agencies (including INGOs [international non-governmental organizations]); the willingness of the technocracy to extend its operations into community-based service delivery; and the willingness of communities and their representatives, the CBOs [community-based organizations] to take on responsibility for the full O&M [operation & management] of their water supply systems" (p. 56).

[9] The idea of Water Warriors or *Jal Yodhas* has been used by the websites and publications of the Centre for Science and Environment and was also in the Training Manual on Local Integrated Water Resource Management (IWRM) in Rajasthan, finalized by the Technical Assistance (TA) team of the European Union–State Partnership Programme (EU-SPP) on Water in Rajasthan and representatives of the Government of Rajasthan in a 3-day workshop in Jaipur in December 2013 (EUSPP, 2014).

water availability *regardless of rainfall.* With the Ground Water Department drilling piezometers in all GPs, monitoring these levels and publishing them on the government website, there was much greater public interest in the competition. But as Gopal Rao said, "The real incentive for us, of course, was that the winner for the Water Warriors competition got Rs. 1 crore!" There were 10 second prizes of Rs. 50 lakhs each and 20 third prizes of Rs. 25 lakhs each, and this was a huge bonus for the GPs in the state.[10]

11.6 WATER SHARES FOR GREATER STAKES

Sita Devi, the Head of the neighboring Anantasagaram Panchayat, noted however that the real motivation for women in all the GPs of the state was the innovation of "water shares," where each household bought a minimum of one water-share, worth Rs. 100, and in case they won the prize, they got a proportionate share in the prize money![11] A part of the "share capital" raised in this manner was used by the GP for Diwali celebrations in case the village did not win the prize and so it was a win–win situation for all of them. But what the "water share" really did was to give each household in the village a stake in improving the water situation—whether it was by reporting leakages in the pipelines or maintaining the wastewater channels (outside their houses) or ensuring that the roof rainwater-harvesting systems were working properly in the school and GP office.

The competition had unleashed the creativity of the local villagers and they had come together to discuss how to win the competition. To reduce agricultural water use, they first replaced their varieties with less-water-using crops (such as Madagascar rice) and started growing new winter crops (like mustard) that used less water. But noting that this still did not stop the falling groundwater levels, they came up with the idea of monitoring individual bore wells—with an incentive. All farmers who maintained their water level of 1 March year-on-year received an incentive of Rs. 5000 from the GP and those who reduced it further received a bonus, in proportion to their reduction: basically, the more you reduced, the more you earned! This was the major reason, Gopal Rao felt, why groundwater levels had begun to rise in the GP.

[10] This idea is taken from the 2001 Clean Village Campaign for rural sanitation in Maharashtra, where the winning GP received Rs. 40 lakhs as total prize money. See Pragmatix Research & Advisory Services and Swayam Shikshan Prayog (2005) for details.

[11] The idea of "shares" in community enterprise is not new, and has been used, for instance, in Joint Forest Management (JFM) to raise contributions for JFM activities (such as forest patrolling, fencing, digging holes, and creating plantations) in places such as Banswada in south Rajasthan where some villages had started raising such "share capital" in the 1990s. There is also the celebrated case of Sukhomajri (Joshi and Seckler, 1982).

11.7 TANKAS TO SHORE UP DOMESTIC SUPPLIES

But for Sita Devi, her source of pride was elsewhere.The increased awareness and interest in water issues also helped her to introduce innovative practices to combat future droughts and summer scarcities, which caused unequal hardship for the women in her village. After her exposure visit to Uddawas village in Rajasthan, she said that she had been very impressed by the tanka system that she had seen there: every house had two tankas (small tanks) of around 10,000L each, one filled by rain water collected from the roof of the house and the other filled by the government piped water supply. During summer scarcities, these two filled tanks were sufficient to provide a family of five with 10L of water for domestic purposes per person for 400 days! She said that she managed to implement this scheme in Brahmasamudram with funds from the government Rural Water Supply Department and today every house has two such tankas.[12] She also organized training on how to use this water and explained: "now we make sure the tankas are kept full from 1 March onwards and so, even if the rains are poor in June, we have enough water for basic domestic needs for more than 12 months!"[13]

She also extended the idea to livestock and so, besides the cleaning and renovation of traditional ponds for livestock use, several large tanks had been built in different parts of the village to fill the cattle troughs that had been built to water the livestock. Gopal Rao was all praise for her actions, saying she had made sure that no animal would die of lack of water even if there were droughts.She also mentioned that all existing drinking water systems had been re-engineered to become multiple-use systems (catering to livestock and water-using activities, including small enterprises such as brick-making, tea-shops and laundries), while all new systems were being designed and built using this concept of 'multiple use'.

11.8 WORKING WITHIN "WATER DISCHARGE LIMITS"

That was not all, Janardhan said, the district administration had imposed "water discharge limits" on each village and town to make sure that surface water flowed down to the rivers. He explained that, noting the earlier problems with upstream water-harvesting reducing water availability downstream

[12] The example of Uddhawas GP is detailed in James et al. (2015b). The provision to support traditional water management systems is contained in Annexure 1 of the National Rural Drinking Water Programme (NRDWP) of the Ministry of Drinking Water and Sanitation, Government of India (GOI, 2010), but has hardly been used by implementing state government departments of Rural Water Supply (RWS) or Public Health Engineering (PHED).

[13] This example is from Bharu GP and Ismailpur village in Sopura GP of Jhunjunu district in Rajasthan. See James et al. (2015b) for details.

(which I was familiar with), the State Water Policy had also stipulated the formulation and enforcement of Water Discharge Limits (WDLs) on rivers and large streams.[14] Based on hydrological assessments of how much water was being used in different stretches of rivers in different times of the years and according to the rainfall, "dynamic" WDLs had been formulated by each district administration, in collaboration with the urban administrations within the district. While scientists had given their findings, the MPs, MLAs, and the bureaucrats had formed a Water Discharge Level Committee (WDLC) to determine what these WLDs should be, based on economic, social and environmental criteria. All watersheds were then given their (dynamic) WDLs, which were notified by the government twice a year, once prior to the monsoon season (based on weather forecasts) to help the farmers plan their kharif cultivation, and another around Diwali time, to help plan the rabi cultivation. Given the strict punishment for violations, he said, villagers themselves would check the website to see whether automated water-level recorders had recorded violations—not just by their village but also by upstream villages. He said the scheme was a success because, after the first set of violating GPs was fined Rs. 25 lakhs each, there were no further violations!

11.9 THE WATER FUND FOR SUSTAINABILITY

When I asked them how they had managed to continue these efforts for 5 whole years, they explained that this was largely due to the increased awareness of the local community about water, and how they could conserve it and actually reduce its use, but there was also a question of finances. Sita Devi explained that the GP had a sizeable Water Fund of around Rs. 1 crore, half of which had come from the annual revenue from the sale of water shares: some people had bought even 2000 shares per person, and had been amply rewarded when their village had won second prize in the state 2 years ago. Also, with the improvements in the water supply system in the village—now providing 24-h piped water supply with individual house meters—the GP were able to raise the water tariffs (since people were now satisfied with the high quality of their water supply).[15] Another part of the Water Fund had been accumulated from

[14] The idea of "water discharge limits" was discussed during the study reported in James et al. (2015a).

[15] That villagers were happy to pay for better water supply services was evidenced in the Tamil Nadu Rural Water Supply Pilot Project in 2004–6 (Nayar and James (2010). Realizing that the noticeable improvements in their household-level water supply was due to better collection, which in turn brought better maintenance of their household piped water systems, these householders not just agreed to pay their own dues, but also volunteered to pay a certain monthly sum for the maintenance of public standposts which, they said, were used by the poor who could not afford to pay.

the annual income from auctioning the rights to common assets, such as the village pond (for fishing) and common lands for tree plantation (of citrus, tamarind, and other horticultural activities by Livelihood Groups), for cultivation of rainfed crops (by landless families), and for livelihood activities (eg, vermiculture, beekeeping, food processing, and packaging and solid waste separation by Livelihood Groups). They now had enough money to replace the entire piped water system in their village, Gopal Rao said, at the end of its lifetime.[16]

11.10 LOCAL VARIATIONS ON A COMMON THEME

Variants of the same basic story were heard from all the 10 states I visited, from arid Rajasthan and Gujarat in the west through Maharashtra and Madhya Pradesh to Bihar, Jharkhand and Orissa in the east and Karnataka, Andhra Pradesh, and Tamil Nadu in the south: they all had enough water to last them through local water scarcities and to cope with multiyear droughts. The basic ingredients were the same, although there were local variations.

- **CCA accounts**: In Bihar, for instance, the state government had helped the communities to open "climate change adaptation" (CCA) bank accounts, where 10% of all contributions to savings accounts were compulsorily deducted—and was made available for households, with a matching contribution from the bank (as a loan), solely for investments in "adaptation" activities. This had been used by farmers to invest in digging tankas, going in for water-saving seed-multiplication units (in polyhouses, using fertigation), "vertical" vegetable cultivation (ie, growing these plants in beds placed in racks that could be stacked up), and mulching of field crops - all of which helped reduce their vulnerability to droughts.[17]
- **Pressure taps and supply lines**: Although there was piped and metered water to all individual houses (as in all the other states) all public taps in rural Tamil Nadu had been fitted with pressure taps, to eliminate leakage, and supply timings had been worked out streetwise with the local community. Schoolchildren were involved, along with Self Help Group women to monitor these public taps, which had reduced leakages from public taps to zero.[18] My memories of dusty and dirty

[16] This was the case, for instance, in the villages in rural Orissa where the NGO called Gram *Vikas* has been working (Banerjee, 2011).

[17] The idea of "CCA accounts" is being discussed in the DfID-supported South Asia Programme on Climate Change Adaptation (2014–2019) being implemented by Oxford Policy Management (OPM) in Afghanistan, Bangladesh, Nepal, Pakistan and India (in the six states of Assam, Bihar, Chhatisgarh, Kerala, Maharashtra, and Orissa).

[18] This initiative is from the Change Management initiative of the Tamil Nadu Rural Water Supply Pilot Project. See Pragmatix Research and Advisory Services (2007) for details.

village streets, with water overflowing from public stand posts with no taps, remained what they were—just memories—as I walked through village after clean village.

- **Water support networks**: In rural Karnataka, Maharashtra, and Andhra Pradesh, the GPs had signed an Memorandum of Understanding (MoU) with individual farmers, whose bore wells had sufficient water during summer, to forego cultivation for that season and supply water to the GP instead, during droughts and summer scarcities[19]—for which they were to be paid well (up to 50% of the value of the foregone crop). This ensured that even during consecutive years of droughts, these wells were able to supply local drinking water to the GPs, often at zero cost, as the farmers were convinced that making water available to their fellow-villagers was in fact an important social contribution they were making (like a "divine duty").

This was indeed a different India than I had left. But I still had nagging doubts about the larger picture, and to address these I decided to talk to the district administration about the details of the institutional and governance support provided.

11.11 THE DISTRICT PICTURE: MODERN SCIENCE MEETS SMART GOVERNANCE

Meeting Ms. Amudha, the District Collector, was a revelation. In addition to the usual traits of an efficient and capable government bureaucrat, she was remarkably well-informed about water resource management. It was the severe 3-year drought of 10 years ago that started the change, she explained, and the Government of India had begun intensive training on water resource management for all its civil service officers, from the Administrative Service to the Forest Service to even the Accounts and Audits Service. This she felt was essential for addressing the rural drinking water crisis, which was highlighted in the media following the failure of rains. Addressing state Finance Department officials, in particular, was important, she felt, since many promising initiatives could not be tried because of the reluctance of these officials - who were more concerned about procedures and had little idea about the overall need, objectives and importance of such initiatives. The course taught her, she said, for the first time the importance of "non-water sectors" in managing water resources:

[19] Personal experience, as part of field work undertaken in a village in Bijapur district for the Karnataka Watershed Development (KAWAD) project in 2001.

Earlier, the Rural Water Supply or Public Health Engineering Department saw its role only as a department responsible for constructing piped water supply systems to distribute water sourced from bore wells or from surface water sources. The sustainability of the source was not a major issue because they felt that water could always be brought, even from distant locations, in case the local source failed. But with our exposure to hydrological modelling mapping and measuring water resources, we realized that water is not an infinite resource and escalating costs of bringing water to villages was not a sustainable solution. But it was when we turned our attention to addressing the problem that we realized that uncontrolled ground water extraction for agriculture was the real culprit for the drying up of local drinking water sources and that we could not address this problem without looking at the larger issue of water resource management – which needed us to work with other departments such as the Agriculture Department, Forest Department, Panchayati Raj Department and even the Highways Department.

The state government had put together a think-tank of smart young civil service officers (of which she was one) and they had worked for almost a year to find a sustainable solution. They had intensive discussions with hydrologists and groundwater experts, engineers, and meteorologists; studied how other countries like Australia and the USA (especially California) had tried to address the problem, and brainstormed about the most effective way to address the issue in the specific context of rural India. "We quickly realized that we could not 'import' ideas from abroad and that we had to innovate and adapt these into locally suitable solutions."

After meeting another 50 collectors across the 10 states, and finding that their accounts were remarkably similar, I started piecing together how they had done it. The solution they had worked out rested on five "pillars."

11.12 THE FIVE PILLARS OF CHANGE

- **Basing planning on basin-level hydrological modeling**: Every district in the state was part of a nation-wide hydrological system, feeding data into the system and receiving locally relevant information for decision-making.[20] All departments had pooled the data from their automatic weather stations and water-level recorders and the state had a huge coordinated network of such data stations.[21] Apart from providing farmers with timely information (through mobile phones) of rainfall and

[20] On the need for and possibilities of hydrological modeling to support decision-making in water resource management in India, see James et al. (2015a).

[21] The 8-year National Hydrology Project of the Ministry of Water Resources, Government of India, supported by the World Bank is aiming to create such a hydrological and meteorological data network.

weather-related information to help them plan their sowing and other agricultural operations[22]—these data also supported the hydrological modeling that was being done by the state-level Water Resources Data Centre.[23] These models were disaggregated at district level but also "nested" within the larger state model (and interstate models in cases of shared rivers and catchments). And these were available at all district headquarters on large-display GIS maps. The modeling helped the planning of all water infrastructure in the state across departments, including how much water should be "reserved" for droughts in various water bodies. But these models also ran simulations for summer scarcities, deficit rainfall years and multiyear droughts, and helped administrators to plan for water-related contingencies in advance (these also ran flood scenarios, in preparation for climate-change-linked "Black Swan" events).[24]

- **Setting and enforcing water discharge levels:** All rivers and streams had water discharge levels, set by a high-level committee of elected representatives and bureaucrats (the District Water Missions) and monitored by the GIS-based monitoring system set up and operated by the Water Resources Department (WRD). These limits were conditional on rainfall and announced before each sowing season, so that GPs could plan their water use. Each GP therefore knew how much surface water had to be released downstream, measured by an automated water-level recorder set at specified points along the water course. Villagers were trained on how to check the readings on the publicly available government website that showed the readings at each point, and managed their own water resources accordingly. The penalties for violations were stringent and therefore respected. The WDLs themselves were reviewed by the WLC every 3 years. GPs planned their local water use to ensure that total water use was within their local WDLs.
- **Supporting water resource planning**: All GP-level water resource plans are coordinated at block and district levels, not just consolidated. That meant that the cropping, livestock and livelihood water uses

[22] The NGO, Water Organizations Trust (WOTR), has already started such a scheme for farmers in parts of Maharashtra (see www.wotr.org), while data-based recommendations on sowing times are already being supplied to farmers by the Government of India through RSL, a private service provider. Personal communication, Shri. R. Tiwary, Assistant District Manager, Uttarakhand Livelihood Support Project, Almora district, Uttarakhand, 29 November 2015.

[23] Although State Water Data Centres have been set up in 14 states under the Hydrology Project Phase II of the Ministry of Water Resources, Government of India, supported by the World Bank (and to be further strengthened during the National Hydrology Project from 2016–2024), an exploration of the potential work that a State Water Data Centre could do is in Kumar and James (2013).

[24] See Taleb (2007).

specified in the GP Water Plans were checked against the block and district Water Resource Plans, using the modeling software, to ensure that the planned water use would stay within their WDLs.[25] For this, the WRD had created a simple format that the GP could access and fill up on mobile phones and send directly to a district-level database through a simple SMS. All GPs, MPs, and MLAs had been trained on using this system and it had been rolled out with the support of Technical Agencies, using staff (and students) from universities, research institutes, government training institutions, and local NGOs. However, there was substantial emphasis given on rejuvenating and reviving traditional water-harvesting and conservation systems, supported by government funds for field activities, capacity building and documentation of these practices from different parts of the state. All these were made available on the website of the newly constituted Water Resource Management Mission (WRM Mission) of each state. In addition to GP-level plans, the district-level water resources planning has also to take into account interbasin transfers, including canals that run across districts or across blocks within a single district, and volumetric discharges from interdistrict rivers and large streams, though these are based on rainfall and demand projections.[26]

- **Encouraging water demand management**: A key part of the solution was the identification of a range of water-saving techniques, ranging from microirrigation, improved cultivation techniques, water-saving crops and output price-incentives for less-water-using crops (including subsidized crop insurance) to soil conservation techniques (including mulching to conserve in situ soil moisture) and water conservation options (such as farm ponds, poly-housing of water bodies to reduce evaporation).[27] These had been prepared in consultation with scientists from ICAR and international research organizations, and selected based on results from experiments done on actual farmers' fields. This was a growing field with new options being continually vetted and added, and all these were also on the website of the WCM. This was an essential part of the "Water Warriors" scheme, which would not have succeeded if reduced water demand had led to lower farmer profits.

[25] Cloud-based data collection and virtual laboratories are mentioned in James et al. (2015a) as part of a suite of new technologies that are currently available.

[26] A methodology for creating district water plans was detailed in Kumar and Bassi (2013), an unpublished report submitted to the TA Team of the EU-SPP.

[27] See EUSPP (2014) for lists of these options.

- **Promoting diversified livelihoods**: Another key part of the solution was the focus on providing alternative livelihoods to local villagers to reduce their traditional dependence on agricultural and livestock production. This was based on the identification of a large number of livelihood activities, based on a value chain analysis, and a gradation depending on their dependence on water, and then the formulation of interdepartmental support packages to ensure that the local communities received the training and awareness to benefit from these. These were both individual- and group-based. Examples individual non-water-based livelihood activity include migration support (providing migrants going to urban areas to work on construction sites, loading, brick-making, etc., with ID-cards [to show police and other authorities if asked], health insurance cards and money transfer services [like Western Union], etc.) and providing training to youth (going beyond the conventional "vocational" training in carpentry, plumbing, and electrical maintenance to computer education, sales and marketing, and ecotourism).[28] Examples of group-based activities included creating opportunities for self-help groups (SHGs) and producer groups (PGs) to add value to agricultural products (eg, by storing, packaging, and processing), to engage in agricultural marketing of both inputs and outputs (through Livelihood Cooperatives), to operate Sanitary and Building Material Marts (buying in bulk and supplying in village clusters) and to collect, sort, package, and export handicrafts and medicinal and aromatic plants.[29]

The basic point about this approach was the explicit recognition that solutions to the water crisis lay *outside* the water sector. What was also refreshing to note was that, not only had each of these been formulated keeping in mind the "incentive" for the stakeholder (ie, the "what is in it for me?" question), especially for local community members, but that these were workable within the government system. It looked as if the use of government officials in the "think tank" (instead of just consultants and experts, as in the "old days") had paid off.

I had come across these five pillars in one way or the other during my earlier stint in India, but they were almost always at pilot scale in

[28] On migration support programs, see Deshingkar et al. (2008). On ecotourism, see www.villageways.org. On livelihood support, in general, see the work of the Integrated Livelihood Support Project by the Uttarakhand Gramya Vikas Samiti (UGVS), supported by the International Fund for Agricultural Development (IFAD), at www.ugvs.org, and the work of the World Bank-supported Uttarakhand Decentralized Watershed Development Project (now in Phase II), locally known as the Gramya project, at www.wmduk.gov.in/UDWDP.html.

[29] See, for instance, the activities of the UGVS at www.ugvs.org.

donor-supported projects and, although there were several "islands of excellence," the challenge was to sustain these beyond the pilot stage and to apply them through the government system at a sufficiently large scale. It looked like they had cracked this problem, but I was still not sure about how far these processes had been "institutionalized." For, in the India that I knew, unless a procedure was enshrined in a "program," with "guidelines" and a "budget," so that it could be implemented by government institutions, structured from national to state to village levels, it would not sustain. Although the Collectors assured me that bringing water to drought-stricken villagers by tankers and trains, and requisitioning agricultural wells of farmers, were in the past now, I still had my doubts about the sustainability of the Five Pillar Approach.

11.13 STATE INSTITUTIONS: VISIONING A NEW REALITY

Apart from meeting secretaries, directors, and other senior bureaucrats at state-level, I also met several elected representatives—a first for me, since I had focused only on bureaucrats during my earlier work in India. In those days, although I had recognized that politicians were the ones that really called the shots in the state—and to whose tunes, bureaucrats had to dance—I had found it difficult to approach them. This time around, I not only had the backing of the Ministry of External Affairs (whose Department of Economic Affairs had cleared my trip and acted as a liaison with state governments concerning my trip) but the exchange program between the elected representatives of Indian states and those of other countries had helped enormously to sensitize these elected representatives of the international interest in the work they had done. Phone calls opened doors and I was received warmly by Chief Ministers, Ministers, MLAs, MPs, and of course, the bureaucrats.

Through the many hours of discussions with elected representatives in the 10 states, I began, once again, to sense a pattern of thought and action: these were politicians who recognized that (1) they had to work across party lines to address the water problem sustainably; (2) there were immediate electoral gains to doing so; and (3) they needed to stay engaged with the issues, challenges, and possible solutions, even as opposition party members, so that they could continue the momentum if and when they were in power.

Bureaucrats, on the other hand, were instrumental in carrying out what the politicians were convinced about—but not longer had to keep them informed about new options and possibilities. Private sector knowledge management agencies had been tasked with keeping not only the general public informed through their website, blogs, local language newspapers, and television channels, but had also had to organize regular monthly presentations, exposure visits, and one-on-one briefings for elected representatives (starting with the cabinet), on new options, technologies and pilot activities in other regions of the country and in other countries. They also had to prepare briefing notes and papers, with the top academicians in the country and outside, for presentations on topics of interest to these elected representatives. In short, there had been a sea change in the awareness of elected representatives and their demand for information on water. And this had generated the necessary reforms, both within the water sector and also in those sectors like agriculture, animal husbandry, forestry, livelihood promotion, and highways and road development, which had in turn improved rural drinking water supply.

As Ms. Sunaina Ibrahim, the Principal Secretary of the PHED in Orissa explained, "only through the presentations and discussions were we able to understand clearly, the link between irrigation for agriculture and domestic water use for humans and livestock. Once we were able to appreciate that the water resources available to any region are finite—limited by the rainfall and the available stocks of ground and surface water—we realized what needed to be done to ensure sustainable rural drinking water supply. We saw that the overall demand for water had to be reduced, and a part of the water saved during good rainfall years had to be made available for domestic water "buffers"—to be used during droughts and other water-scarce periods, just like in the past.[30] Once that realization was there, the rest was easy—we are after all very familiar with how to work with the government system and achieve our objectives!"

Mr. Kasinathan, MLA from Bellary in Karnataka, however, sought to restore the "balance of power" in the perennial tussle between politicians and bureaucrats on attribution: "Only because we MLAs realized that the IAS officers and engineers had to be better educated on water resource management we allowed them to go for training programs and exposure visits, and that is how they have got the knowledge today. On their own, they would not have done any of this—like they had not done in the 77 years since Independence! We politicians realized this and so facilitated their learning. We are all getting the benefit of this learning now."

The major state-level changes that had been brought about following this "enlightenment on water resource management" were twofold, as I

[30] For a discussion of "buffers" in earlier times in India, and "deficits," see James (2012).

could make out across all the 10 states: (1) legal reform to provide a supporting framework for community co-management of water resources, and (2) institutional integration within the government, to facilitate rapid and coordinated action that was both evidence-backed and science-based.

11.14 SUPPORTING COMMUNITY MANAGEMENT THROUGH LAWS

As Dr. Sadashivan Nahule, MP from Maharashtra's Sangli district explained to me, the laws of the state of Maharashtra had vested ownership and control over the state's water resources in the hands of the government since the passing of the Northern India Canal and Drainage Act of 1873 and the Easement Act of 1883. The subsequent edifice of laws, rules, and regulations built upon this key principle even after political independence in 1947 (James et al., 2014). While this was largely done in order to recoup money from farmers for the large investments made by the British Government in India for the dams, weirs, and canal network created to increase agricultural production, it also effectively removed age-old institutions for the community-management of water (such as the neerakattis in Andhra Pradesh and Karnataka and the *kohlis* of Himachal Pradesh and Uttarakhand) and complex arrangements for water-sharing in irrigation systems such as the phad system of Maharashtra and the tank-based cultivation of peninsular India (see CSE, 1997). Recognizing that this biased legal heritage was a major bottleneck in the way of effectively involving local communities in the management of water resources, the government had instituted national and state commissions to study these laws and modify them to provide incentives for community management. The key legislation in this regard was the national-level Water Resources (Management and Regulation) Act of 2019, which was the basis for all subsequent legislation at national and state levels.

It was under these new Acts that most of the field-level reforms had been undertaken, including the hydrological modeling, the coordinated action by different departments at GP level to reduce water demand management and improve rural livelihoods simultaneously. "The coordinated action by MLAs of all parties made this possible" he said, echoing what MLAs of other states had said. Indeed, the political democracy in India had led from the front on this occasion, with India's future at stake, instead of the usual obstructionist opposition to all ruling party proposals. This was yet another lesson to take away.

11.15 INTEGRATING THE WATER INSTITUTIONS

Since all government actions are routed through an institutionalized hierarchy, it was essential that appropriate changes were made so that the "Indian elephant" changed direction.[31] But, once informed, aware and united as to the purpose, Indian politicians and bureaucrats were equal to the task and had unleashed a slew of reform measures, including the following:

- **Water Policies:** the new National Water Policy had been quickly followed by similar State Water Policies, emphasizing three specific aims of water resource management in India:

  Rationalize water use to allocate water to the most desirable uses, on the basis of both hydrological modeling (to find out how much water was available in each basin, sub-basin and watershed—and their interconnections) and consultations with stakeholders (to decide the objectives of water resource management, based on social, economic, environmental, political and other locally relevant factors). It was from these consultations that the Water Discharge Limits were developed for all the watersheds and basins in the country. Unlike the Australian and South African cases, which were studied, the Indian model suggested a two-part allocation process: First, discussions in a high-level committee (eg, the State or District Water Mission) formed within a unit (eg, a district or a state) and comprising key members from among elected representatives (MPs, MLAs, Mayors, CEOs, and Additional CEOs of Zilla Parishads), senior rural and urban administrators (eg, District Collectors, Municipal Commissioners) from within the unit, and representatives of all major stakeholders, including industrial owners, farmer's and merchant's bodies, and NGOs. Supported by a consultant team that provided all the necessary hydrological and other information, this committee fixed WDLs for the region—ie, how much water would be released through the different surface water sources that flowed into the downstream hydrological unit (eg, a stream or river flowing into the neighboring district). Second, stakeholder representatives then

[31] Apart from being symbolic of India, the elephant has also been put to good use as a metaphor for progress and change in India. In James (2008), for instance, the key idea is to compare Indian development to the slow and stately walk of an elephant, occasionally prodded to run by a well-designed program, policy or initiative, but to achieve the Millennium Development Goals by 2015, this elephant would have to fly.

agreed on how the total amount of water available in the unit would be shared between them.[32]

Reduce crop water use, while maintaining rural livelihoods, to create water savings for creating water "buffers" in high rainfall years (to use during water stress periods such as droughts and summer scarcities). Again, this was done on the basis of the hydrological modeling but coupled with innovative approaches from all sectors whose actions affected water use (eg, Panchayati Raj, Agriculture and Highways Departments; Banks and other financial institutions; and Rural Livelihood Missions). The old Prime Minister's Krishi Sinchayee Yojana (PMKSY) had been decentralized to districts, working in close coordination with the WRM Missions of different states, and block-level staff were now implementing its key provision of reducing surface water and groundwater use and creating water buffers to store water in high rainfall periods for use in rain-deficit periods.

Reverse groundwater mining: Given the rampant overextraction of groundwater in earlier decades, the policy unequivocally stated that mining had to be not just eliminated but reversed, with success being measured in *increases* in groundwater levels. The policy, however, noted that this required a carrot-and-stick approach, and hence authorized stringent action on overextraction (based on scientifically determined annual targets) and incentives for achieving and exceeding the targets. Supportive national government provisions included the incentive Water Warriors scheme (implemented at state-level), privately provided Knowledge Management Centres (supplying different options to reduce groundwater draft) and the nationwide publicity given to groundwater levels and multiple use systems in national print and television media to ensured that issues were constantly receiving public attention. Apart from these, each state was free to use other means to achieve the policy goals.

- **Mission Mode:** Since the non-achievement of objectives within the 5-year window specified in the National Policy attracted penal budgetary provisions in terms of allowances, salaries and perks of elected representatives and all government servants, many states had achieved their

[32] This process not only eliminated the paralysis of decision-making in large groups as in South Africa (see James (2006) on an early assessment of problems of operationalizing Catchment Management Agencies), and the distrust of the negotiation process in Australia caused by lack of credibility of scientific data (see Syme et al., 2012, and Box 5 in James et al., 2015a for two case studies).

targets ahead of time (for which there was a bonus). All states had created a Water Resources Management Mission (WRM Mission), under a bureaucrat with the rank of an Additional Chief Secretary (senior to most heads of individual departments, who were at most of the rank of Principal Secretary), to report directly to the Chief Secretary and Chief Minister, and empowered to coordinate the activities of key departments (such as Water Resources, Agriculture, Rural Water Supply/Public Health Engineering, Rural Development, and Panchayati Raj) to achieve the goals of the national Water Policy. The WRM Mission had access to a Water Resource Knowledge Management Centre (Water KMC), which was a donor-supported initiative (initially) to provide a high-quality consultant team for 5 years to support state-level decision-making on water resource management. After the expiry of this term, on the basis of its usefulness, all state governments had continued their Water KMCs with 50% government funding and 50% funding from Indian corporates (from the additional 0.5% of corporate profits that was now required to go solely for WRM, as part of their corporate social responsibility). The WRM Missions had been set up under the new WR (M&R) Act of 2019 and hence violations of its orders and charges, including unacceptable service delivery standards for domestic water supply, were now punishable in a court of law.

A key preoccupation was how best to implement, in practice, the idea of co-management of water by local communities and government, and the WRM Missions and the District Water Missions were the direct result of these deliberations.

11.16 CONCLUSIONS

The remarkable achievements of India's water sector had several inter-related elements, including policy, institutional, and legal reforms coupled with far-sighted programming and support mechanisms. But to me the main take away is that the pace of reform can be significantly enhanced when the onus of delivery is laid clearly on the elected representatives of the country and its individual states. Once the citizens realize that it is basically their elected representatives that can resolve their water problems satisfactorily, they can put pressure on these representatives to deliver what they need: sustainable water for all domestic uses, 24 × 7. And this can set off a domino effect, where the pressure from citizens—through a variety of means, including social media—results in politicians leaning on bureaucrats to find satisfactory answers, and the latter sourcing information from experts and resource persons, analyzing the best way to institutionalize it within government programs and policies and then ensuring that the system implemented these to achieve the desired results.

But I did realize that the Indian phenomenon had benefited from some "pump priming."The actual success were critically dependent on the earlier work by committed NGOs, academics, and experts (the *Imagine India Coalition*), working with the (social) media to draft a blueprint of possible action *before* citizens' pressure was used to galvanize elected representatives into action. There was also a lot of organized and planned awareness generation and "immersion" of local, state and national politicians by the Coalition to prepare the ground, so that elected representatives were able to respond. This was the real backbone of the success and other countries seeking to emulate the Indian example will do well to learn this lesson too.

REFERENCES

Banerjee, A., Duflo, E., 2011. Poor Economics: A Radical Rethinking of the Way to Fight Global Poverty. PublicAffairs, New York.

CSE, 1997. Dying Wisdom: Rise, fall and potential of India's traditional water harvesting systems. Centre for Science and Environment, New Delhi.

Deshingkar, P., Khandelwal, R., Farrington, J., 2008. Support for Migrant Workers: The Missing Link in India's Development. Natural Resource Perspectives. Overseas Development Institute, London.

EUSPP, 2014. Local Integrated Water Resource Management: Training Manual. Jaipur: European Union State Partnership Programme Technical Assistance Team and State Water Resources Planning Department, Government of Rajasthan.

GOI, 2010. National Rural Drinking Water Supply Programme. Department of Drinking Water Supply, New Delhi. Ministry of Rural Development, Government of India.

GoO, 2010. Guidelines for construction of check dams. [pdf] Bhubaneshwar: Department of Water Resources, Government of Orissa. Available at http://www.dowrorissa.gov.in/DownLoads/Guidelines/CheckDams/Guidelines%20for%20Constn%20of%20Check%20Dams%20082010.pdf.

James, A.J., 2006. Institutional challenges for water resources management: India and South Africa. In: Nair, P., Kumar, D. (Eds.), Water Management: Concepts and Cases. ICFAI University Press, Hyderabad.

James, A.J., November 2008. Giving wings to the elephant: towards creative governance for urban sanitation. In: Paper Presented at the International Symposium on Urban Sanitation for the Poor. IRC International Water and Sanitation Centre, Delft. At: http://www.irc.nl/page/45285.

James, A.J., 2011. India: Lessons for Rural Water Supply: Assessing Progress towards Sustainable Service Delivery. IRC International Water and Sanitation Centre and Delhi: iMaCS, The Hague. http://www.irc.nl/page/67087.

James, A.J., 2012. Demystifying climate change: policy implication for agricultural and rural development for drought adaptation in semi-arid India. In: Singh, S., Mohanakumar, S. (Eds.), Climate Change: An Asian Perspective. Hyderabad and Guwahati: Rawat Publications, Jaipur, New Delhi, Bangalore.

James, A.J., Kumar, M.D., Mandavkar, Y., Suresh, V., Snehalatha, M., 2014. India Case Study on Water Tenure [Report]. Submitted to the Food and Agriculture Organization, Rome.

James, A.J., Kumar, M.D., Batchelor, J., Batchelor, C., Bassi, N., Choudhary, N., Gandhi, D., Syme, G., Milne, G., Kumar, P., 2015a. Catchment Assessment and Management Planning for Watershed Management. World Bank, Washington DC.

James, A.J., Rathore, M.S., Nishchal, H., Bikram, K., Jha, N., 2015b. Monitoring and Evaluation of the EC-assisted State Partnership Programme (Rajasthan), Rajasthan Study 5, European Commission- State Partnership Programme in Rajasthan, Special Studies Series. ICF International & Institute of Development Studies, London & Jaipur.

Joshi, D., Seckler, D., 1982. Sukhomajri: water management in India. Bulletin of the Atomic Scientists. March. 26–30.

Kumar, M.D., Bassi, N., 2013. Water Accounting Study of Luni River Basin and District IWRM Planning for Pali [Report]. Technical Assistance Team of the European Union – State Partnership Programme on Water in Rajasthan, Jaipur (Unpublished).

Kumar, M.D., James, A.J., 2013. A Note on the State Water Data Centre for Rajasthan. [Note]. Technical Assistance Team of the European Union – State Partnership Programme on Water in Rajasthan, Jaipur (Unpublished).

Michael, A.M., 2009. Irrigation: Theory and Practice. Vikas Publishing House, NOIDA. P. 69.

Nayar, V., James, A.J., 2010. Policy insights on user charges from a rural water supply project: a counter-intuitive view from South India. International Journal of Water Resources Development 26 (3), 403–421.

Pragmatix Research & Advisory Services and Swayam Shikshan Prayog, 2005. Inventive Villagers: Innovative Approaches to Total Sanitation in Maharashtra. [Report]. Water and Sanitation Program – South Asia, New Delhi (Unpublished).

Pragmatix Research & Advisory Services, 2007. Quantified Participatory Assessment of the Impact of Change Management Training to Engineers. [Report]. Tamil Nadu Water Supply and Drainage Board, Chennai (Unpublished).

Syme, G., Reddy, V.R., Pavelik, P., Croke, B., Ranjan, R., 2012. Confronting scale in watershed development in India. Hydrogeology Journal 20 (5), 985–993.

Taleb, N.N., 2007. The Black Swan: The Impact of the Highly Improbable. Random House, New York.

WHiRL, 2004. Using watershed development to protect and improve domestic water supplies. [pdf]. Briefing Note. Water, Households and Rural Livelihoods project. Natural Resources International Limited, Greenwich. Available at http://projects.nri.org/wss-iwrm/Reports/India_tank_briefing.pdf.

CHAPTER 12

Building Resilient Rural Water Systems Under Uncertainties

M.D. Kumar
Institute for Resource Analysis and Policy, Hyderabad, India

A.J. James
Institute of Development Studies, Jaipur, India

Y. Kabir
UNICEF Field Office, Mumbai, Maharashtra, India

12.1 RESILIENT MULTIPLE-USE WATER SYSTEMS: SUMMARY OF EVIDENCE FROM DIFFERENT TYPES OF SYSTEMS

From the empirical evidence available from case studies presented in this collection, we can summarize the conditions under which the sustainability of rural water systems are threatened vis-à-vis their effectiveness in meeting the multiple water needs of rural people, as follows:

- *High interannual variability in hydrological conditions in an area*, which change both water availability and water demands (especially agricultural water demands);
- *Interseasonal variability in water availability and marked differences in seasonal demand patterns* between competing water uses such as agriculture, fisheries, and drinking water;
- *Dependence on water-resource systems, whose characteristics are not known*, ie, the stock is not properly quantified, and to which access cannot be restricted, and the number of users and uses are either not defined or unknown; and
- *The hydraulic linkages between water systems*, with one causing negative physical externalities on the other.

These, however, also help to identify factors that can improve the resilience of these systems, given uncertainties.

First: systems with enough provisions for multiannual storage of water in order to store water in years of high inflows and operational rules that provide for carryover storage would be resilient to droughts. Many water-resource

Rural Water Systems for Multiple Uses and Livelihood Security
ISBN 978-0-12-804132-1
http://dx.doi.org/10.1016/B978-0-12-804132-1.00012-3

systems that are designed to carry inflows having poor dependability. However, the water demands of the sector or sectors for which such systems are designed are such that all that water gets used up in a single year, keeping no carryover storage. For instance, irrigation tanks normally dry up by the end of summer, even in wet years, due to the excessive evaporation and cultivation of a second or third crop. However, if there are restrictions on cultivation during summer months, there would be some buffer storage in the tank to take care of the needs in the event of delayed arrival of monsoon or an imminent monsoon failure in the subsequent year.

In the case of large water-resource systems, multiannual storage can be achieved differently: either by storing water in the reservoirs as buffer for low rainfall years or by supplying the excess water to the command areas for irrigation during the rainy season (part of which in turn could be available as recharge to the shallow aquifers from irrigation return flows). The beneficiary farmers in the command would be able to use this water during droughts as well for irrigation when canal water releases reduce. However, in such cases, the cropping pattern in the command area should be such that it can withstand excess water releases. Paddy is the most suitable crop for allowing recharge through water spreading. The non-beneficiary farmers and non-farmers on the other hand can use the recharged water in good as well as bad years. A slightly modified version of this strategy was found in the case of Sardar Sarovar project.

Second: systems with flexible-use patterns are likely to survive. It is idealistic to think of a multiple-use water system, which would perform all its functions at the same level across seasons and years, when hydrological uncertainties are high. The performance of the systems would change, depending on the year (ie, whether a wet year, a dry year, or a normal year) and the season. A resilient system would have a low degree of fluctuation in the overall performance levels across years. Such a system would ensure that when certain water uses suffer (due to insufficient flows or excess flows) from the natural system, certain other uses thrive. For instance, in the case of tank systems in semiarid regions, it is quite common that during dry years, when the tank does not have much water for irrigation of the second crop or sometimes even supplementary irrigation of kharif cultivation, the tank bed is used for cultivation of fodder crops, tapping the soil moisture, and whereby (especially in large tanks) the biomass needs of a large number of cattle-rearing households are met. However, in a good rainfall year, this particular use would be absent and, instead, fish production from the tank, along with irrigated crop production would go up exploiting the large increase in water-spread area of the tanks (and to an extent the increase in volume of water).

The way in which the multiple-use services of wetlands change across the years in accordance with changes in hydrological conditions also depends on the general agro-climatic conditions of the regions that these wetlands serve. In high-rainfall regions, the value realized per unit volume of water from the tanks for irrigation during wet years will not be as high as that in dry years. This is because paddy is the most preferred crop in high-rainfall regions and marginal return from the use of water in irrigation would be low during the monsoon season, with the crop's survival and yield dependent mostly on the direct precipitation falling on the crop land. At the same time, since the volume of water which can be put to use will be large, the reduction in economic value of water use may be offset to some extent.

Similarly, the paddy wetlands in flood-prone, high-rainfall areas (like in eastern UP and north Bihar) can function as multiple-use systems. The inundated areas of flood-prone basins receive micronutrients, planktons, and fingerlings of local fish varieties, along with the floodwaters. Such areas can actually serve as fertile ground for raising fish and shrimp. If paddy is raised in such areas, along with fish and shrimp production, the system would be able to survive in high- as well as low-rainfall years. In the event of excessively high rainfall resulting in floods, the farmers could get some fish and some leafy biomass from paddy (for feeding the animals) during the monsoon season, if the crop gets destroyed in the floodwaters. The residual soil moisture, however, will ensure high winter crop production, with minimum dependence on groundwater for irrigation. On the other hand, in years of normal or below normal monsoon, the production of monsoon paddy could be good, with no fish or shrimp production, but farmers will have to depend on other sources of water (groundwater) for the second crop.

12.2 PLANNING OF RESILIENT RURAL WATER SYSTEMS FOR MULTIPLE USES

Fatal errors made in the past in planning of rural water systems for household needs include the underestimation of both the demand for water by households and the neglect of the importance of water allocation in ensuring adequate amount of water for these high-priority needs (given the competition this sector faces from other competitive water-use sectors). In most agro-climatic regions in India, irrigation takes a large volume of water in terms of consumptive water use. Furthermore, in most regions, from hot and hyperarid Rajasthan to the cold and humid northeastern mountainous region, due to monsoon weather patterns, the water availability in the natural system changes drastically across seasons—with the largest proportion of the natural flows of water during the monsoon season and the lowest amount of water during the summer season. Against this, the demand for water for irrigation increases from the lowest during the monsoon season to the highest during the summer season.

Though the intensity of irrigation water demand (volumetric demand for irrigation water per unit of arable land) can change significantly across agro-climates—with the highest intensity of irrigation water demand being in the hot and hyperarid regions and the lowest being in the cold and humid regions—in accordance with the availability of arable land, and thus the variation in demand across seasons is very sharp. The highest intensity of irrigation demand is always during the summer season. In contrast to the demand pattern in the irrigation sector, the demand for water for household needs remains more or less the same throughout the year, though marginal increase in the demand for water is generally seen during the summer season.

The unique challenge, posed by the mismatch between water demand and natural water availability in the hydrological system, in managing water supplies for household needs can be overcome through two measures. Firstly, by tapping and storing a sufficient amount of natural flows of water in the hydrological system in storage reservoirs and earmarking this for the high-priority demand during the lean season (when the flows in the natural system dwindle). Secondly, by using the water which is naturally available in the hydrological system for household needs during the lean season (which means banning irrigation water use during that season). There are many regions in India where sufficient water could be found in the hydrological system to meet household water needs including some of the most water-scarce regions of the world, as household water needs are quite meager in comparison to the total amount of water that the basins in these regions yield. A good example is the Luni river basin in western Rajasthan.

Against the total renewable water resources (the sum of annual dependable runoff and the average annual groundwater recharge) of around 2606 million cubic metres (MCM) per annum (2011–12) that the basin produces at a dependability of 75%, the total water consumed by domestic sector, including urban consumption, was only 110 MCM, whereas the water consumed in irrigation was 2404 MCM (Kumar and Bassi, 2013). A large proportion of this water is in the form of annual groundwater recharge from rainfall and the balance fraction only is runoff. Nevertheless, such arid and hyperarid regions also experience high interannual variability in weather parameters, especially rainfall and temperature. The changes in rainfall also induce disproportionately higher changes in stream flows and thus, there could be some dry years, during which a sufficient amount of surface water will not be available to meet even household water needs. In any case, the river basins in semiarid, medium- to high-rainfall regions receive runoff which is far more than adequate to meet the domestic requirement, provided reservoirs and water conveyance infrastructure are built to store and take the water to the places of demand.

However, the second option is socially unviable, largely because the livelihoods of a majority of the population in most rural areas of India are heavily dependent on farming. Agriculture requires the artificial application of water even in the coldest region of India, ie, the northeast, though for the 2 months of summer, in February and March. The number of months for which cropping requires irrigation can be as high as 10, in the western parts of Rajasthan and northern parts of Gujarat. Hence, all the water available in the natural system would get appropriated to meet the irrigation demand. This tragedy is common in areas where the natural endowment of water is in the aquifers, because groundwater is still an open-access resource in terms of ownership rights, and there is no control over the abstraction of groundwater by individual farmers.

There are some alluvial areas in India where water would still be available for meeting household needs during summer months in spite of excessive demand for water for irrigation. For instance, high-yielding tube wells can be sustained in the aquifers underlying Indo-Gangetic plains in most parts; permeability is often in the range of 10–60 m/d (metre per day) and specific yield (the drainable porosity) varies from 5 to 20%, making it highly productive (MacDonald et al., 2015). This is due to the abundant groundwater stock in the basin (based on GOI, 1999). But the chemical quality of water poses challenges—with excessively high levels of fluoride, TDS, chloride, nitrate and, in some cases, even arsenic. The problems of poor chemical quality of groundwater are encountered even in the groundwater-rich alluvial belt of the Indo-Gangetic plains, with high TDS in most parts and arsenic in the eastern Gangetic plains (MacDonald et al., 2015)[1]. Hence, groundwater-based schemes would not be a viable option in most situations.

[1] High salinity and elevated arsenic concentrations exist in parts of the Indo-Gangetic basin, limiting the usefulness of the groundwater resource. Saline water predominates near to the coast in the Bengal Delta and is also a major concern in the Middle Ganges and Upper Ganges (Southern Punjab, Haryana, and parts of Uttar Pradesh). Arsenic severely impacts the development of shallow groundwater in the fluvial influenced deltaic area of the Bengal Basin (MacDonald et al., 2015).

Therefore, the only option to ensure water security for household needs is to build reservoirs to capture the runoff in the basins and supply it through a large distribution network. However, in low-rainfall arid regions, water would have to be imported even during dry years, when the runoff sharply declines. Since surface water resources are a common property and state-owned, the government can exercise control over its uses. However, this control is also slowly disappearing with many state governments promoting decentralized water harvesting, most of which is happening in the upper catchments of reservoirs. There are many recent examples from semiarid parts of Gujarat (Kumar et al., 2008) and Rajasthan (Ray and Bijarnia, 2006; Times of India, 2012), where reservoirs (Gujarat) and lakes (Rajasthan) earmarked for drinking water supply are drying up.

Such issues withstanding, it is not an easy task to build reservoirs and water conveyance infrastructure to create dependable water supply systems. A major issue is of finding funds for the high amount of capital investment that is usually required. As the household water demand in rural areas is very small and highly dispersed and water has to be supplied through pipelines to protect the supplied water from all kinds of contamination, the cost of production and supply of water per unit volume would be high. Past experience shows that public water utilities generally shy away from water supply technologies that are capital-intensive, and instead prefer cheaper options such as hand pumps and bore wells based on local groundwater resources. This is despite the poor dependability of these systems, as discussed earlier, their short life and high failure rates during peak summer months. The real issue is that prior to taking investment decisions, proper evaluation of various water supply options is rarely undertaken in a way that factors in the real cost of ensuring sustainability of water supplies. The ultimate result is that low-cost options, which have a short life and which are unsustainable, gain weightage in the process of selection of water supply technologies, instead of cost-effective ones.

Even with large reservoirs and regional water transfers to meet domestic and productive water needs of rural households, local drinking water schemes based on groundwater could still be used in normal rainfall years to meet water needs during the kharif season and to an extent during the winter season, as sufficient water would be available in the aquifers during these two seasons. Such a model has been successfully tried in Saurashtra and Kachchh regions of Gujarat in western India, where water from single-village-based water supply schemes (such as bore wells) and regional water supply schemes (based on medium and minor reservoirs, catering to a group of villages) are used to meet the needs during the monsoon months, and the water from the Narmada-based pipeline scheme is used during peak summer months when the local water sources dry up. This results in optimum utilization of local and exogenous water resources.

12.3 MANAGEMENT OF RURAL WATER SYSTEMS FOR MULTIPLE USES

Four key physical and technical issues challenge the management of rural water supply systems for multiple uses.

12.3.1 Quantity of Water

In a rural area, where a system is designed to provide water for the domestic needs of the households, a system for both domestic and productive needs, then the quantum of water required would be much larger. Water for livestock would constitute a major share of that demand. In a hot and dry area, a fully grown buffalo weighing 450–500 kg would need to drink as much as 75–80 L of water a day, and an indigenous cow, 45–50 L of water a day. If a small vegetable garden of 10 sq. km has to be irrigated during summer months, it would require 75–120 L of water a day (if evapo-transpiration (ET) is assumed to be 7.5–12 mm).

Adequate quantities of water will have to be supplied from the public system to meet these needs, especially during summer months when the local informal water sources (such as ponds, tanks, springs, and farm wells) dry up. The resource, which is being tapped to supply water, should have an adequate amount of water—which could be a big challenge for schemes based on groundwater sources in hard rock regions. Such problems, however, should not affect surface systems that are primarily designed for irrigation (such as large reservoir-based schemes) as the additional demand for household needs would be a very small fraction of the total demand. Nevertheless, for systems such as village irrigation tanks, evaporation from the reservoir during summer months could be exceptionally high owing to the low depth of the water column, and water may not be available for domestic uses during the lean season.

12.3.2 Quality of Water

If a system is designed for low-value uses such as irrigation, and if water from the same system has to be supplied for meeting high-priority demands such as domestic water supply, the quality of water will have to be improved to make it free from physical, chemical, and microbial contaminants. This does not mean the entire water from the irrigation system will have to be of potable quality, as that may not be economically feasible. Instead, a subsystem, which would tap water from the irrigation system and treat it to drinkable standards, will have to be created. The treatment system should have an intermediate storage system; a filter to remove physical and microbial impurities; and water distribution pipelines. Such systems would make economic sense, especially if the local area does not have any reliable source, or if the available water requires expensive treatment to make it potable. A good example is the Indira Gandhi Nahar Project (IGNP), of which several rural and urban drinking water supply schemes have been built to supply to hot and arid areas where groundwater is highly saline and where there is no other reliable source of surface water. Hence, it would make good economic sense to invest in simple treatment systems to make the irrigation water from canals suitable for domestic water supply.

12.3.3 Modifying System Operations to Address Multiple Water Needs

Supplying water to meet certain uses that the system was initially not designed to cater to would require changes in system operation. Water supply schedules for irrigation and household needs generally do not match: Most irrigation systems supply water either during kharif season or during winter and do not supply water during summer months, whereas domestic water supply will have to be year round. Hence, additional storage facilities will have to be created to store sufficient canal water to tide over summer scarcity. Such operational issues would come up even for irrigation systems that are primarily designed for field crops, such as paddy and wheat, but where farmers in the command area want to raise horticultural crops instead (or in addition). Horticultural crops such as coconut, arecanut, mango, pomegranate, and banana require watering during summer months, and therefore irrigation schedules will have to change—and, accordingly, the reservoir operating rules also will have to change.

To give an example, a major reason for the failure or poor performance of many gravity irrigation systems in the state of Kerala in peninsular India was the change in land use, which the state has experienced since the 1970s[2]. While surface irrigation systems based on gravity flow were designed for

[2] Since 1970, the total cropped area in Kerala remained more or less stagnant, while the area under rice cultivation plummeted by about 526,000 ha—a 60% drop between 1975 and 2003. Concomitantly, there has been a pronounced "coconut and rubber boom." That is, coconut area increased by 106% between 1955 and 2000; thereafter it, however, stabilized. Likewise, rubber area expanded in the state during the period from 1955 to 2000 by 627%. Other crops that gained substantial coverage over the same period include arecanut and plantain (Kumar, 2005).

irrigating wetland paddy, their long gestation periods (30–40 years), resulted in large-scale shifts in cropping pattern (from paddy to plantation crops such as coconut and banana), and major changes in the landscape (through raising the land levels for proper drainage of rainwater during monsoon) in the interim, and these gravity-based systems were not able to supply water to these plantations because of the elevation difference and lack of flexibility in the operational rules.

12.3.4 Technical Infrastructure

If a rural drinking water supply system has to be retrofitted to meet the demands of water for livestock and kitchen gardens, then it has to be done in such a way that the retrofitted system takes water to the backyards of dwellings (in the case of kitchen gardens), or extra feeder pipes from the drinking water stand post and water troughs for animals. In certain cases, schemes to augment village drinking water sources are planned to make them capable of supplying water during summer months. Generally, recharge schemes are built in the vicinity of the water supply well so that the recharged water augments the output from the well.

In the case of irrigation canals, either operational rules would have to be changed to keep water running in the canals perennially or water storage systems have to be built to store water for summer months along with pumps, filters, and pipelines to take treated water to the village. Also, steps will have to be constructed in the canal to enable village people to satisfy other uses, for example, for women to go down to bathe and wash clothes.

In the case of village irrigation tanks in semiarid and arid regions which dry up during summer months (owing to low depth and high-evaporation rates), a small portion of the tank bed can be dug out so that water remains in the dug-out portion (with a deep water column even during the lean season) and is earmarked for domestic uses. In order to prevent contamination of water in the tank, it would also be necessary to create a baffle wall as a gated structure separating the dug-out portion from the rest of the tank water-spread area.

12.4 INSTITUTIONS AND POLICIES

In an era of growing water scarcity, managing water supply for multiple needs is largely about managing water allocation (Saleth and Dinar, 1999; Kumar, 2010). But water allocation becomes a management issue only when the resource is scarce in qualitative and quantitative terms and across space (regionally and locally) and time (across years and seasons). Hence, institutions are required for allocating water when the resource is *scarce* (Kumar, 2010).

Planning of water supply services in most developing countries is sectoral and separate line agencies look after different service sectors such as urban water supply, rural drinking water supply, irrigation, and industrial water supply. Each of these agencies accordingly designs "single-use" systems, ie, to supply water only to its concerned sector. There are, similarly, separate departments for livestock development and development of inland fisheries, although water supply for these sectors is not within their purview. Unfortunately, the latter agencies have little coordination with either the irrigation or drinking water supply departments to ensure water availability for their activities. Thus, while multisectoral water allocation of water is likely to become an increasingly important issue as water resources become scarce, today there are no agencies at regional or basin levels to allocate water across different sectors, as per the allocation priority fixed as per State or National Water Policies.

With growing regional competition for water owing to scarcity (ie, when total water demand for all competitive uses exceeds total freshwater supplies), it has become a common phenomenon in India to reallocate water from public reservoirs for urban use at the cost of irrigation (Kumar, 2010). Such diversion, however, increases the pressure on groundwater resources as farmers drill more wells to water their crops, often causing excessive depletion of groundwater and drying up of drinking water supply sources in rural areas. In certain emergency situations, drinking water also gets priority in water allocation from public reservoirs, when District Collectors pass orders to "reserve" water in public reservoirs. However, under such circumstances, water for fisheries, livestock, and vegetable gardening in rural areas gets the lowest priority. Public water utilities generally supply the bare minimum, hardly ever exceeding 40 L per capita per day.

What is therefore needed is a basin-wide allocation of water across sectors, guided by a variety of social, economic, and environmental goals, and based on certain principles (though this would require resolving issues related to water rights, treated separately below). That said, the goals and principles have to be reflected in the water policy of the state or the county. The goals ultimately chosen to base water allocation would depend on the values attached to each goal.

There are three major challenges involved in water allocation. These are described in the following subsections.

12.4.1 Deciding on the High-Priority Water Uses in a Basin or Region

Generally, human drinking and domestic water use receive the highest priority in all water allocation decisions, as they serve social goals. Thereafter, the remaining water is allocated to other uses in the order in which priorities are decided. These priorities ought to be determined by the socioeconomic profile of the region in question and the various goals to be pursued—and therefore should be based on larger basin-wide consultation of the primary stakeholders of water, especially the direct users. Whether livestock water demand, fisheries water demand, or water demand for backyard cultivation would be a high priority use should be determined by the occupational profile of the people living in the rural areas of the region, and preferred (future) occupations of the local communities.

If a region is socially and economically well-developed, communities might attach high value to environmental management (Rosegrant et al., 1999). Allocation of water amongst the users within each sector shall then be decided on the allocation principles. Equity, efficiency/productivity, and sustainability are some of the principles for intrasectoral water allocation.

12.4.2 Quantification of Water Needs

The practice of estimating water demands for livestock using a simplistic norm of 30 L per capita (followed by some Indian states) ought to be replaced by a more scientific approach that considers the number of animals under different livestock categories, climatic conditions, and the average body weight of the animals. Domestic water demand can also be estimated more rationally (based on climatic conditions and the overall socioeconomic conditions of the people) in terms of access to piped water supply, access to improved sanitation, etc. Per capita water demand for domestic needs would be high, for instance, in areas with improved access to water supply and sanitation systems. There has to be a basin-level regulatory institution that decides this allocation in volumetric terms, and enforces it—and here all the lessons from past experience will have to be applied (see, for instance, Singh et al., 2013).

12.4.3 Ownership Rights Over Water

While the state can claim ownership rights over surface water appropriated from natural catchments using reservoirs and diversion structures, allocating water from aquifers poses a major challenge. In most developing countries, ownership rights to groundwater are not clearly defined in volumetric terms. In India, de jure rights to groundwater are not clear. De facto, the rights to use groundwater are attached to land ownership rights under English Common Law (Singh, 1995), which is still followed in the country (Kumar, 2007). This is the case with India's neighbors, namely, Pakistan, Nepal, Sri Lanka, and Bangladesh. Therefore, if the groundwater is brought under the ambit of basin-wide water allocation, there could be legal problems in enforcing the allocation norms. Hence, it is important that only surface water is considered in arriving at (volumetric) allocation decisions for high-priority needs, otherwise, serious compromises will have to be made in achieving basin allocation goals.

Though the ownership rights over surface water are clear in India, the government is not able to regulate the development and use of surface water generated in catchments—largely because of the increasing decentralization in development and management of water resources over the past couple of decades or so. Water is a state subject in India but several state government agencies (at the state, district, and Panchayat level) are engaged in the development of surface water resources, through building of major, medium, and minor irrigation projects; implementing water-harvesting and recharge schemes; and watershed development activities. In addition, NGOs and community-based organizations are also engaged in water-harvesting and watershed development activities. Currently, the activities of these agencies are not coordinated at the level of catchments and there is an institutional vacuum in terms of regulation of surface water use.

However, once the allocation is decided on volumetric criteria, it should be the responsibility of this regulatory institution to ensure that the line agencies concerned adhere to this allocation plan while planning water development schemes to meet their sectoral needs, as this plan becomes statutory. However the allocation norm—the basis for dividing the water across sectors—should not be static and should instead vary according to the basin inflows (the renewable groundwater resources and annual runoff). The allocation norm becomes more important when the renewable water resources in the basin become extremely limited. Under such circumstances, the high-priority demands for water will have to be met and lower priority demands may have to be compromised in order to protect the former. In wet years, there would be more water available for meeting the low-valued uses. However, such values are not always decided on the basis of simple economic criteria (economic value of the benefits produced from water use), and can instead be social or environmental.

Apart from higher-order institutions for water allocation, the local institutions for running the schemes are also essential to ensure sustainable water supply. In recent decades, institutions of local communities have gained prominence in the management of rural water supplies in developing countries (Hutchings et al., 2015). However, village, community-based organizations are created today using a stereotypical approach focusing on representation from community members, but with no major support for technical issues, and monitoring and finances. Recent research shows that for community management models to be successful and sustainable, a number of internal and external factors have to act in tandem (Hutchings et al., 2015):

- Extensive and long-term support for local community institutions.
- Broader socioeconomic development of the community.

Since "community management" is viewed either as a cheap service delivery model or a goal in itself, there is a growing tendency to go for simple water systems that operate at the scale of hamlets and individual villages, which the communities are willing to manage. This has to give way to a pragmatic approach to community management, in which the role of external agencies is considered to be extensive, involving serious commitment in terms of investment in support institutions.

Along with institutions, policies will also have to change. The policy of supplying 40 L of water for domestic uses per capita and below, especially in drought-prone areas, in spite of the norm from National Rural Drinking Water Programme for providing a minimum of 55 L per capita of water per day in all rural areas and the goal of achieving 70 lpcd of water supply in all rural households by 2022, will have to change (NRDWP, 2013). The justification for this, that managing a larger quantum of water in drought-prone areas would be extremely difficult, is however, frivolous, as illustrated earlier for Luni river basin in Rajasthan. Furthermore, the norm does not consider the water needs of livestock while, in many semiarid and arid regions with growing scarcity of water for agriculture, livestock keeping and dairy farming are emerging as very important economic activities. Farmers find it easier to manage (green and dry) fodder for animals than managing water for growing conventional field crops. However, this changing phenomenon has not influenced the water supply policies of public water utilities.

As illustrated through a multicountry analysis (chapter: Water, Human Development, Inclusive Growth, and Poverty Alleviation: International Perspectives), in countries which are at low levels of social development and economic growth, returns on investment in improving water security through water infrastructure, institutions, and policies can be high in terms of reduced hunger, improved human development, higher per capita GDP, reduced income inequity, and poverty reduction. Therefore, no matter how much it costs, it is important to use the best available technologies to build water systems in rural areas, capable of supplying adequate quantities of good-quality water for drinking, cooking, and other domestic needs as well as for irrigation and livestock throughout the year including during natural disasters. While community-based organizations are increasingly becoming the norm as a viable institutional model for management of rural water systems, an enabling policy framework is needed to ensure greater investment in support institutions. Institutional reforms to achieve this need to include the creation of new institutions for water management at various levels from the state to river basins, watersheds, and village Panchayats, as also for water pricing and water rights (Kumar, 2010). Such support institutions can nurture community-based organizations, through the involvement of civil society organizations, as shown by the experience of WASMO in Gujarat, where such institutions proved to be not only critical to, but also effective in facilitating, community management at the local level.

REFERENCES

Government of India, 1999. Integrated Water Resource Development: A Plan for Action. Government of India, New Delhi. Report of the National Commission on Integrated Water Resources Development-Volume I.

Hutchings, P., Chan, M.Y., Cuadrado, L., Ezbhake, F., Mesa, B., Tamekawa, C., Franceys, R., 2015. A systematic review of success factors in the community management of rural water supplies over the past 30 years. Water Policy 17, 963–983.

Kumar, B.M., 2005. Land use in Kerala: changing scenarios and shifting paradigms. Journal of Tropical Agriculture 42 (1–2), 1–12.

Kumar, M.D., 2007. Groundwater Management in India: Physical, Institutional and Policy Alternatives. Sage Publications, New Delhi.

Kumar, M.D., 2010. Managing Water in River Basins: Hydrology, Economics, and Institutions. Oxford University Press, New Delhi.

Kumar, M.D., Patel, A.R., Ravindranath, R., Singh, O.P., 2008. Chasing a mirage: water harvesting and artificial recharge in naturally water-scarce regions. Economic and Political Weekly 43 (35), 61–71.

Kumar, M.D., Bassi, N., 2013. Water Accounting Study of Luni River Basin and District IWRM Planning for Pali. Final Report Submitted to the Technical Assistance Team of the EUSPP, Jaipur, Rajasthan.

MacDonald, A.M., Bonsor, H.C., Taylor, R., Shamsudduha, M., Burgess, W.G., Ahmed, K.M., Mukherjee, A., Zahid, A., Lapworth, D., Gopal, K., Rao, M.S., Moench, M., Bricker, S.H., Yadav, S.K., Satyal, Y., Smith, L., Dixit, A., Bell, R., van Steenbergen, F., Basharat, M., Gohar, M.S., Tucker, J., Calow, R.C., Maurice, L., 2015. Groundwater Resources in the Indo-Gangetic Basin: Resilience to Climate Change and Abstraction. British Geological Survey, Nottingham, UK.

National Rural Drinking Water Programme, 2013. Movement towards Ensuring People's Drinking Water Security in Rural India-Framework for Implementation (Updated 2013). National Rural Drinking Water Programme, Ministry of Drinking Water and Sanitation, Government of India, New Delhi.
Ray, S., Bijarnia, M., 2006. Upstream versus Downstream: groundwater management and rainwater harvesting. Economic and Political Weekly 41 (23), 2375–2383.
Rosegrant, M.W., Ringler, C., Gerpacio, R.V., 1999. Water and land resources and global food supply. In: Proceedings of the 23rd International Conference of Agricultural Economics, Sacramento, California.
Saleth, R.M., Dinar, A., 1999. Water Challenge and Institutional Responses (A Cross-Country Perspective). Policy Research Working Paper Series 2045. World Bank, Washington, DC.
Singh, K., 1995. Cooperative property rights as instruments for managing groundwater. In: Moench, M. (Ed.), Groundwater Law: The Growing Debate. Monograph, VIKSAT-Natural Heritage Institute, Ahmedabad.
Singh, S., Reddy, V.R., Batchelor, C., Marothia, D.K., James, A.J., Rathore, M.S., 2013. Regulating Water Demand and Use in Rajasthan, European Union State Partnership Programme, Special Studies Series Rajasthan 4, Institute of Development Studies and IPE Global, Jaipur, Rajasthan, March 2013.
Times of India, 2012. Damned Ramgarh Still Dry. Times of India, Jaipur.

INDEX

'*Note:* Page numbers followed by "f" indicate figures, "t" indicate tables.'

D

L

M

N

S

Printed in the United States
By Bookmasters